SpringerBriefs in Electrical and Computer Engineering

Computational Electromagnetics

Series editors

K. J. Vinoy, Bangalore, India
Rakesh Mohan Jha (Late), Bangalore, India

More information about this series at http://www.springer.com/series/13885

Hema Singh · Mausumi Dutta
P.S. Neethu

Probe Suppression in Conformal Phased Array

Hema Singh
Centre for Electromagnetics (CEM)
CSIR-National Aerospace Laboratories
Bangalore, Karnataka
India

P.S. Neethu
Centre for Electromagnetics (CEM)
CSIR-National Aerospace Laboratories
Bangalore, Karnataka
India

Mausumi Dutta
Centre for Electromagnetics (CEM)
CSIR-National Aerospace Laboratories
Bangalore, Karnataka
India

ISSN 2191-8112   ISSN 2191-8120   (electronic)
SpringerBriefs in Electrical and Computer Engineering
ISSN 2365-6239   ISSN 2365-6247   (electronic)
SpringerBriefs in Computational Electromagnetics
ISBN 978-981-10-2271-5   ISBN 978-981-10-2272-2   (eBook)
DOI 10.1007/978-981-10-2272-2

Library of Congress Control Number: 2016947382

Printed on acid-free paper

This Springer imprint is published by Springer Nature
The registered company is Springer Science+Business Media Singapore Pte Ltd.

To Late Dr. R.M. Jha

Preface

In a given signal scenario, the phased array is expected to radiate in such a way that sufficient gain is maintained towards the desired directions, with no energy transmitted towards the hostile radars. This can be achieved with the help of an adaptive array processing, which involves an efficient adaptive algorithm. Further other design parameters of antenna array, platform effect, mutual coupling between the antenna elements affects the array performance. The surface over which antenna array is mounted affects the radiation and scattering characteristics of phased array.

This book describes the probe suppression in a cylindrical microstrip patch and dipole array. The effect of both conducting and dielectric cylinder is discussed. For such non-planar geometry, the radiation pattern synthesis is done by transforming element pattern using Euler rotation matrix. The optimal weights are calculated using the modified improved LMS algorithm. The adapted and quiescent antenna array patterns are generated for a given signal environment consisting of both desired and probing sources. It is shown through several illustrations that the array mounted over a cylinder along with an efficient adaptive algorithm is able to cater to the impinging signals, whether desired or hostile sources. The adapted pattern maintains a sufficient gain towards the desired sources with accurate and deep nulls towards each of the probing sources. This book includes the detailed analytical description of the active cancellation of probing sources by phased arrays mounted on non-planar conducting and dielectric surfaces, algorithm for weight adaptation towards the generation of antenna pattern, and numerous simulation results.

Bangalore, India

Hema Singh
Mausumi Dutta
P.S. Neethu

Acknowledgements

We would like to thank Mr. Jitendra J. Jadhav, Director, CSIR-National Aerospace Laboratories, Bangalore for the permission to write this SpringerBrief.

We would also like to acknowledge valuable suggestions from our colleagues at the Centre for Electromagnetics, and their invaluable support during the course of writing this book. We would like to thank the project staff at the Centre for Electromagnetics, for their consistent support during the preparation of this manuscript.

But without the concerted support and encouragement of our Springer editorial contacts, it would not have been possible to bring out this book within such a short span of time. We very much appreciate the continued support extended by Springer.

About the Book

This book considers a cylindrical phased array with microstrip patch antenna elements and half-wavelength dipole antenna elements. The effect of platform and mutual coupling effect is included in the analysis. The non-planar geometry is tackled by using Euler's transformation towards the calculation of array manifold. Results are presented for both conducting and dielectric cylinders. The optimal weights obtained are used to generate adapted pattern according to a given signal scenario. It is shown that the array along with adaptive algorithm is able to cater to an arbitrary signal environment even when the platform effect and mutual coupling is taken into account. This book provides a step-by-step approach for analyzing the probe suppression in non-planar geometry. Its detailed illustrations and analysis will be useful for graduate and research students, scientists and engineers working in the area of phased arrays, low observables and stealth technology.

Contents

About the Authors

Hema Singh is currently working as Principal Scientist in *Centre for Electromagnetics* of CSIR-National Aerospace Laboratories (NAL), Bangalore, India. Before NAL, she was Lecturer in EEE, BITS, Pilani, India during 2001–2004. She obtained her Ph.D. degree in Electronics Engineering from IIT-BHU, Varanasi India in 2000. Her active area of research is Computational Electromagnetics for Aerospace Applications. More specifically, the topics she has contributed to are GTD/UTD, EM analysis of propagation in an indoor environment, phased arrays, conformal antennas, Radar Cross Section (RCS) studies including active RCS reduction. She has received the Best Woman Scientist Award at CSIR-NAL, Bangalore for period 2007–2008 for her contribution in area of phased antenna array, adaptive arrays and active RCS reduction. Dr. Singh has co-authored nine books, one book chapter and over 200 scientific research papers and technical reports.

Mausumi Dutta obtained M.Tech. in RF & Microwave Engineering from RV College of Engineering, Visvesvaraya Technological University Bangalore. She obtained her B.Tech. (ECE) degree in 2012 from ICFAI University, Tripura. She has worked on radar cross-section based studies and probe suppression in phased arrays.

P.S. Neethu obtained B.Tech. (ECE) and M.Tech. in Communication Engineering from University of Calicut, Kerela, India. She is a Project Scientist at the Centre for Electromagnetics (CEM) of CSIR-National Aerospace Laboratories, Bangalore, where she works on probe suppression in phased arrays on planar and non-planar surfaces, and RCS studies for aerospace vehicles.

Symbols

α	First rotation Euler angle
β	Second rotation Euler angle
γ	Third rotation Euler angle
δ_{0m}	Kronecker delta function
ε_o	Permittivity of free space
ε_r	Relative permittivity
η	Intrinsic impedance
θ	Elevation angle in global Cartesian coordinate system
$\tilde{\theta}$	Elevation angle in local Cartesian coordinate system
λ	Wavelength of the incident wave
μ	Permeability
μ_s	step-size
ξ	Angle between any two adjacent elements
σ	Conductivity
ϕ	Azimuth angle in global Cartesian coordinate system
$\tilde{\phi}$	Azimuth angle in local Cartesian coordinate system
Φ	Angle between x axis and the last element
ω	Angular frequency
a	Radius of the cylinder
a_n	nth element pattern
$a_{n\theta}$	nth element pattern in θ direction
$a_{n\tilde{\theta}}$	nth element pattern in $\tilde{\theta}$ direction of local Cartesian coordinate system
$a_{n\phi}$	nth element pattern in ϕ direction
$a_{n\tilde{\phi}}$	nth element pattern in $\tilde{\phi}$ direction of local Cartesian coordinate system
a_{nx}	nth element pattern in x direction
$a_{n\tilde{x}}$	nth element pattern in $\tilde{x}$ direction
a_{ny}	nth element pattern in y direction
$a_{n\tilde{y}}$	nth element pattern in $\tilde{y}$ direction
a_{nz}	nth element pattern in z direction
$a_{n\tilde{z}}$	nth element pattern in $\tilde{z}$ direction

b	Distance of dipole from the centre of cylinder
B	Distance from centre of cylinder to observation point
C_i	Cosine integral
d	Inter-element spacing
e	Thermal noise
E_θ^t	Electric field pattern in θ direction
f	Operating frequency
F	Far-field array pattern response
F_θ	Far-field array pattern response in θ direction
F_Φ	Far-field array pattern response in Φ direction
$H_m(\cdot)$	mth order Hankel function of first kind
i_p	Probing signal
I	Identity matrix
I_c	Current
j	Number of probing signal
$J_n(\cdot)$	nth order Bessel function of first kind
k	Number of snap shots
k_o	Wave number
l	Dipole length
m	Number of desired signal
N	Number of antenna elements
P	Projection operator
R	Euler rotation matrix
$\tilde{R}$	Received signal correction matrix
R_{mutual}	Real part of mutual impedance
R_{self}	Real part of self-impedance
s_d	Desired signal
S	Steering vector
S_i	Sine integral
S_o	Desired signal steering vector
S_p	Probing signal steering vector
u_θ	Unit vector in θ direction
$u_{\tilde{\theta}}$	Unit vector in $\tilde{\theta}$ direction in local coordinate system
u_Φ	Unit vector in Φ direction
$u_{\tilde{\phi}}$	Unit vector in $\tilde{\phi}$ direction in local coordinate system
u_x	Unit vector in x direction
$u_{\tilde{x}}$	Unit vector in $\tilde{x}$ direction in local coordinate system
u_y	Unit vector in y direction
$u_{\tilde{y}}$	Unit vector in $\tilde{y}$ direction in local coordinate system
u_z	Unit vector in z direction
$u_{\tilde{z}}$	Unit vector in $\tilde{z}$ direction in local coordinate system
$U_n(\cdot)$	nth order Hankel function of second kind
v	Unit vector from origin to observation point
w	Weight vector

x_n	Received signal of nth antenna element
X_{self}	Imaginary part of self-impedance
X_{mutual}	Imaginary part of mutual impedance
Z	Impedance
(x, y, z)	Global Cartesian coordinates
$(\tilde{x}, \tilde{y}, \tilde{z})$	Local Cartesian coordinates
∇	Gradient vector
$(.)^H$	Hermitian (transpose conjugate)
$(\cdot)^T$	Transpose

List of Figures

Probe Suppression in Conformal Phased Array

Abstract

The probe suppression capability of phased arrays has applications in stealth technology. The suppression of hostile radar sources attempting to probe the phased array mounted on an aerospace structure facilitates the low observability. The modified improved LMS algorithm is known to be an efficient algorithm for active cancellation of probing in phased arrays. This motivates its usage in conformal adaptive array processing. This book considers a cylindrical phased array with microstrip patch antenna elements and half-wavelength dipole antenna elements. The effect of platform and mutual coupling effect is included in the analysis. The non-planar geometry is tackled by using Euler's transformation towards the calculation of array manifold. Results are presented for both conducting and dielectric cylinder. The optimal weights obtained are used to generate adapted pattern according to a given signal scenario. It is shown that an array along with adaptive algorithm is able to cater to an arbitrary signal environment even when the platform effect and mutual coupling is taken into account.

This book provides a step-by-step approach for analyzing the probe suppression in non-planar geometry. Its detailed illustrations and analysis would be a useful text for the graduate and research students, scientists, and engineers working in the area of phased arrays, low-observables and stealth technology.

Keywords Conformal phased array · Probe suppression · Modified improved LMS algorithm · Adapted pattern · Nulls

Probe Suppression in Conformal Phased Array

1 Introduction

When the antenna elements are mounted over a non-planar platform, the radiation and scattering characteristics are very different from that of planar surface. In planar antenna array the alignment of elements is along a specific axis, in contrast the orientation of antenna elements in non-planar array depends on the varying surface curvature. The element pattern is polarization dependent and hence depends on the individual element pattern apart from the geometry of the array. The analysis of non-planar arrays requires transformation of antenna pattern from local coordinate system to global coordinate system. This may be carried out using methods like Euler angle rotation, geometric algebra, etc. The design and analysis of non-planar array is complex and has constraints such as cross polarization, mutual coupling effect and limited operational bandwidth. However, such arrays are known to have reduced radar cross section (RCS), less aerodynamic drag and greater angular coverage. In open-domain methods such as alternative projection method (Wang et al. 2008; Wang and He 2010) and iterative least squares based synthesis method (Vaskelainen 1997) have been proposed for the pattern synthesis of non-planar arrays.

Moreover, the antenna elements placed close to each other results in mutual coupling effect. This degrades the array performance both in terms of radiation and radar cross section. Thus, it is required that radiation pattern analysis and RCS estimation of phased array should include both mutual coupling and platform effect. The design parameters of phased array mounted on the surface should be optimized so as to have reduced scattering and optimum radiation characteristics.

One of the important issues faced by antenna designers and researchers in the design of antenna array systems mounted on platform is the current that flows over the surface, substantially affecting the radiation pattern of antenna array. Consequently, to ensure control over the antenna performance it is critically important to determine the effect of the platform over which the antenna system is

© The Author(s) 2017
H. Singh et al., *Probe Suppression in Conformal Phased Array*,
SpringerBriefs in Computational Electromagnetics,
DOI 10.1007/978-981-10-2272-2_1

placed. In most of the aerospace structures, the platform is more or less cylindrical in shape. This allows the usage of the formulation for the finite cylinder towards an approximate solution. The far-field radiation pattern of dipole antenna placed near an infinite conducting cylinder has been derived using the principle of reciprocity (Carter 1943). The corresponding analysis for finite conducting cylinder was reported by (Kuehl 1961). The portions of the infinite cylinder above and below the cylindrical section considered are ignored with the assumption that the current on the remaining portion of the cylinder remains unchanged. The far-field of this unchanged portion of the current is added to the dipole field to arrive at the total far-field due to dipole antenna placed near a finite cylinder.

This book focuses on phased array such as dipole array, microstrip patch array, placed on a right circular cylinder. The modified improved LMS algorithm (Singh and Jha 2013) is used for weight adaptation so as to determine the array pattern for a signal environment Signal environment consisting of multiple narrowband radar sources. The Euler's rotation is used for transformation of local coordinates to global coordinates towards the extraction of antenna element pattern for cylindrical surface. Multiple desired and probing sources are assumed to impinge the cylindrical array. Results are shown for both conducting and dielectric cylinder. The radiation pattern of dipole antenna placed over a cylindrical surface is computed and validated against the results available in open domain. If multiple antennas share the same ground plane, surface currents can cause unwanted coupling between them. Here, the effect of mutual coupling on the radiation pattern of dipole array is taken into account. The probe suppression is demonstrated for various signal scenarios consisting of multiple desired and probing sources.

2 Steering Vector of Conformal Phased Array

In order to generate the antenna array pattern for a given signal environment , the first step is to determine the array manifold or steering vector.

In non-planar array, the steering vector may be expressed in a global coordinate system (Karimzadeh et al. 2011), as

$$S(\theta, \phi) = \sum_{n=1}^{N} a_n(\theta, \phi) \, \exp(j \, k_o \, r_n \cdot v) \tag{1}$$

where, $a_n(\theta, \phi)$ represents the nth element pattern in the global cartesian coordinate system, $k_o = 2\pi/\lambda$ is the propagation constant, λ is wavelength of the impinging signal. $r_n = [x_n, y_n, z_n]$ is position vector from the origin to the centre of the nth element and $v = [\sin\theta\cos\phi, \sin\theta\sin\phi, \cos\phi]^T$ is unit radial vector from the coordinate origin to the observation point.

The unit vector pointing in the direction (θ, ϕ) of the global coordinate system is given by,

$$(x, y, z) \cdot (r, \theta, \phi) = \sin \theta \, \cos \phi \, x + \sin \theta \, \sin \phi \, y + \cos \phi \, z \qquad (2)$$

Thus, (2) may be rewritten as

$$S(\theta, \phi) = \sum_{n=1}^{N} a_n(\theta, \phi) \, e^{jk_o (\sin \theta \, \cos \phi \, x_n + \sin \theta \, \sin \phi \, y_n + \cos \phi \, z_n)} \qquad (3)$$

2.1 Euler Rotation Method

The radiation pattern of an individual element in a non-planar array is not same as that of an isolated element pattern. In non-planar array such as cylindrical array, the element pattern significantly depends on the element position and the polarization. The normal to each antenna element placed over the non-planar surface points in different directions. In order to bring each surface normal in same direction, the transformation is required.

The transformation from one coordinate system to the other may be carried out using different methods. One of the ways is the Euler rotation, which is used for describing the relationship between two Cartesian-coordinate systems having same origin but different orientations. Here Euler rotation is used to transform from local to global coordinate system and vice versa. The transforming process is carried out in terms of direction cosine matrix (Burger 1995).

Here two right-handed Cartesian and spherical coordinate systems with a common origin is considered, in which the observation point is defined in space as (x, y, z) and (r, θ, ϕ). For a regular 3-D non-planar surface such as cylindrical, conical, etc. only two rotations are sufficient for transformation. The third rotation is in general used for irregular and composite non-planar surfaces.

When microstrip antenna or a dipole is placed over non-planar surface such as cylindrical surface, the element pattern needs to be transformed. The location of each array element on the curved surface is defined in local coordinate system. Since the element pattern is defined in the global coordinates only, a spatial rotation transformation such as Euler rotation is used. The first step in Euler rotation is the formation of Euler transformation matrix. This matrix is obtained by rotating the coordinate axes.

In this book, three successive rotations are carried out for z, y, and z axes. As shown in Fig. 1, first rotation is made about z-axis. The x- and y-axes are rotated by an angle α, such that $[x, y, z] \Rightarrow [x', y', z']$ or $[r, \theta, \phi] \Rightarrow [r', \theta', \phi']$. Then x' and z' axes are rotated by an angle β keeping y'-axis fixed, resulting in $[x', y', z'] \Rightarrow [x'', y'', z'']$ or $[r, \theta, \phi] \Rightarrow [r'', \theta'', \phi'']$. In third rotation, the axes are rotated by angle γ w.r.t. z'' axis. Thus one gets $[x'', y'', z''] \Rightarrow [x''', y''', z''']$ or $[r, \theta, \phi] \Rightarrow [r''', \theta''', \phi''']$. Here, α, β, γ are the Euler angles of rotation.

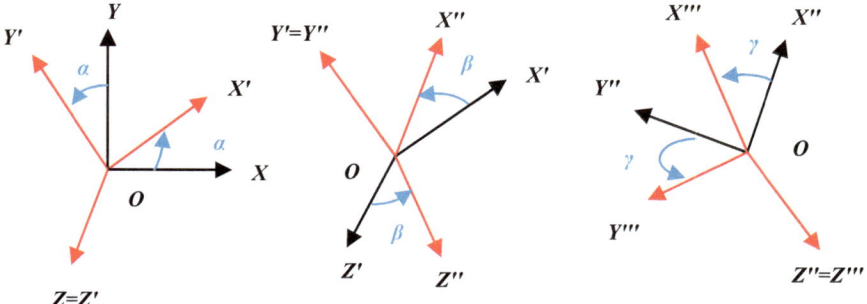

Fig. 1 Three successive Euler rotations with angles α, β, γ

The unit vector before rotation is expressed as

$$P(x, y, z) = x(1, 0, 0) + y(0, 1, 0) + z(0, 0, 1) \tag{4}$$

After first rotation by an angle of α about z-axis, the transformation matrix is obtained as

$$\tilde{P}(x', y', z') = x(\cos \alpha, \ \sin \alpha, \ 0) + y(-\sin \alpha, \ \cos \alpha, \ 0) + z(0, \ 0, \ 1) \tag{5}$$

In matrix form, it can be expressed as

$$\begin{bmatrix} x' \\ y' \\ z' \end{bmatrix} = \begin{bmatrix} \cos \alpha & \sin \alpha & 0 \\ -\sin \alpha & \cos \alpha & 0 \\ 0 & 0 & 1 \end{bmatrix}^{T} \cdot \begin{bmatrix} x \\ y \\ z \end{bmatrix} \tag{6}$$

where rotation matrix corresponding to angle α about z-axis is referred to as *Euler transformation rotation matrix* (ETRM), and is given by (Milligan 1999)

$$R(\alpha) = \begin{bmatrix} \cos \alpha & \sin \alpha & 0 \\ -\sin \alpha & \cos \alpha & 0 \\ 0 & 0 & 1 \end{bmatrix} \tag{7}$$

Similarly, the Euler transformation rotation matrices corresponding to angle β about y-axis and angle γ about z-axis are expressed as

$$R(\beta) = \begin{bmatrix} \cos \beta & 0 & -\sin \beta \\ 0 & 1 & 0 \\ \sin \beta & 0 & \cos \beta \end{bmatrix} \tag{8}$$

$$R(\gamma) = \begin{bmatrix} \cos \gamma & \sin \gamma & 0 \\ -\sin \gamma & \cos \gamma & 0 \\ 0 & 0 & 1 \end{bmatrix} \tag{9}$$

The overall transformation matrix is obtained using (7), (8), (9),

$$
\begin{aligned}
R(\alpha,\ \beta, \gamma) &= R(\gamma)\, R(\beta)\, R(\alpha)\\[4pt]
&= \begin{bmatrix} \cos\gamma & \sin\gamma & 0 \\ -\sin\gamma & \cos\gamma & 0 \\ 0 & 0 & 1 \end{bmatrix} \cdot \begin{bmatrix} \cos\beta & 0 & -\sin\beta \\ 0 & 1 & 0 \\ \sin\beta & 0 & \cos\beta \end{bmatrix} \cdot \begin{bmatrix} \cos\alpha & \sin\alpha & 0 \\ -\sin\alpha & \cos\alpha & 0 \\ 0 & 0 & 1 \end{bmatrix}\\[4pt]
&= \begin{bmatrix} R_{11} & R_{12} & R_{13} \\ R_{21} & R_{22} & R_{23} \\ R_{31} & R_{32} & R_{33} \end{bmatrix}
\end{aligned} \tag{10}
$$

where

$$R_{11} = \cos\alpha\ \cos\beta\ \cos\gamma - \sin\alpha\ \sin\gamma \tag{10a}$$

$$R_{12} = \sin\alpha\ \cos\beta\ \cos\gamma + \cos\alpha\ \sin\gamma \tag{10b}$$

$$R_{13} = -\sin\beta\ \cos\gamma \tag{10c}$$

$$R_{21} = -\cos\alpha\ \cos\beta\ \sin\gamma - \sin\alpha\ \cos\gamma \tag{10d}$$

$$R_{22} = \cos\alpha\ \cos\gamma - \sin\alpha\ \cos\beta\ \sin\gamma \tag{10e}$$

$$R_{23} = \sin\beta\ \sin\gamma \tag{10f}$$

$$R_{31} = \sin\beta\ \cos\alpha \tag{10g}$$

$$R_{32} = \sin\alpha\ \sin\beta \tag{10h}$$

$$R_{33} = \cos\beta \tag{10i}$$

The unit vector pointing in the direction of (θ,ϕ) in global coordinates is expressed as

$$[x \quad y \quad z] = [\sin\theta\ \cos\phi \quad \sin\theta\ \sin\phi \quad \cos\theta] \tag{11}$$

Next, transformation from global coordinate $(\theta,\ \phi)$ to local coordinate $(\tilde{\theta},\ \tilde{\phi})$ is given by

$$
\begin{aligned}
[\tilde{x} \quad \tilde{y} \quad \tilde{z}]^{T} &= R(\alpha,\ \beta,\ \gamma)[x \quad y \quad z]^{T}\\
&= R(\alpha,\ \beta,\ \gamma)[\sin\theta\ \cos\phi \quad \sin\theta\ \sin\phi \quad \cos\theta]^{T}
\end{aligned} \tag{12}
$$

The corresponding elevation and azimuth angles associated with the antenna elements in local coordinates $(\tilde{\theta}, \tilde{\phi})$ are extracted as

$$\tilde{\theta} = a\cos(\tilde{z}); \quad \tilde{\phi} = a\tan\left(\frac{\tilde{y}}{\tilde{x}}\right) \tag{13}$$

2.2 Element Pattern Transformation

In case of conformal phased array, the extracted elevation and azimuth angles, expressed in (13) are substituted in the expression for the radiation pattern of antenna element.

Thus, the pattern of nth array element in spherical local coordinate system may be expressed as

$$a_n(\tilde{\theta}, \tilde{\phi}) = a_{n\tilde{\theta}}(\tilde{\theta}, \tilde{\phi})u_{\tilde{\theta}}(\tilde{\theta}, \tilde{\phi}) + a_{n\tilde{\phi}}(\tilde{\theta}, \tilde{\phi})u_{\tilde{\phi}}(\tilde{\theta}, \tilde{\phi}) \tag{14}$$

where $u_{\tilde{\theta}}(\tilde{\theta}, \tilde{\phi})$ and $u_{\tilde{\phi}}(\tilde{\theta}, \tilde{\phi})$ are unit vectors in $\tilde{\theta}$ and $\tilde{\phi}$ directions respectively, given by

$$u_{\tilde{\theta}}(\tilde{\theta}, \tilde{\phi}) = \cos\tilde{\theta}\cos\tilde{\phi} \cdot u_{\tilde{x}} + \cos\tilde{\theta}\sin\tilde{\phi} \cdot u_{\tilde{y}} - \sin\tilde{\theta} \cdot u_{\tilde{z}} \tag{15}$$

$$u_{\tilde{\phi}}(\tilde{\theta}, \tilde{\phi}) = -\sin\tilde{\phi} \cdot u_{\tilde{x}} + \cos\tilde{\phi} \cdot u_{\tilde{y}} \tag{16}$$

Substituting the value of $u_{\tilde{\theta}}(\tilde{\theta}, \tilde{\phi})$ and $u_{\tilde{\phi}}(\tilde{\theta}, \tilde{\phi})$ in (14) the element pattern $a_n(\tilde{\theta}, \tilde{\phi})$ can be expressed in Cartesian local coordinate system as

$$a_n(\tilde{\theta}, \tilde{\phi}) = a_{n\tilde{x}}u_{\tilde{x}} + a_{n\tilde{y}}u_{\tilde{y}} + a_{n\tilde{z}}u_{\tilde{z}} \tag{17}$$

where in $u_{\tilde{x}}$, $u_{\tilde{y}}$ and $u_{\tilde{z}}$ are unit vectors in the $\tilde{x}, \tilde{y}, \tilde{z}$ directions of local Cartesian coordinate system respectively, and

$$a_{n\tilde{x}} = a_{n\tilde{\theta}}(\tilde{\theta}, \tilde{\phi})\cos\tilde{\theta}_n\cos\tilde{\phi}_n - a_{n\tilde{\phi}}(\tilde{\theta}, \tilde{\phi})\sin\tilde{\phi}_n \tag{18}$$

$$a_{n\tilde{y}} = a_{n\tilde{\theta}}(\tilde{\theta}, \tilde{\phi})\cos\tilde{\theta}_n\sin\tilde{\phi}_n + a_{n\tilde{\phi}}(\tilde{\theta}, \tilde{\phi})\cos\tilde{\phi}_n \tag{19}$$

$$a_{n\tilde{z}} = -a_{n\tilde{\theta}}(\tilde{\theta}, \tilde{\phi})\sin\tilde{\theta}_n \tag{20}$$

Next the transformation of element pattern from local Cartesian coordinates to global Cartesian coordinate is done using inverse Euler matrix, i.e.

$$[a_{nx} \quad a_{ny} \quad a_{nz}]^T = R(\alpha, \beta, \gamma)^{-1}[a_{n\tilde{x}} \quad a_{n\tilde{y}} \quad a_{n\tilde{z}}]^T \tag{21}$$

The overall element pattern in global Cartesian coordinates is expressed as

$$a_n(\theta, \phi) = a_{nx}u_x + a_{ny}u_y + a_{nz}u_z \tag{22}$$

in which u_x, u_y and u_z are unit vectors in x, y and z directions respectively. Likewise the element pattern in the array global spherical coordinate systems is given by

$$a_n(\theta, \phi) = a_{n\theta}(\theta, \phi)u_\theta(\theta, \phi) + a_{n\phi}(\theta, \phi)u_\phi(\theta, \phi) \tag{23}$$

where

$$u_\theta(\theta, \phi) = \cos\theta \cos\phi \cdot u_x + \cos\theta \sin\phi \cdot u_y - \sin\theta \cdot u_z \tag{24}$$

$$u_\phi(\theta, \phi) = -\sin\phi \cdot u_x + \cos\phi \cdot u_y \tag{25}$$

Thus (23) can be rewritten as

$$a_n(\theta, \phi) = \{a_{n\theta} \cos\theta \cos\phi - a_{n\phi} \sin\phi\} u_x \\ + \{a_{n\theta} \cos\theta \sin\phi + a_{n\phi} \cos\phi\} u_y + \{-a_{n\theta} \sin\theta\} u_z \tag{26}$$

Comparison of (23) and (26), the element pattern components are extracted as

$$a_{n\theta}(\theta, \phi) = \frac{-a_{nz}}{\sin\theta} = \frac{a_{nx} \cos\phi + a_{ny} \sin\phi}{\cos\theta} \tag{27}$$

$$a_{n\phi}(\theta, \phi) = -a_{nX} \sin\phi + a_{nY} \cos\phi \tag{28}$$

Thus the steps involved in transformation of element pattern can be summarized as follows:

Step 1   Calculate the Euler angles α, β and γ

Step 2   Determine the Euler rotation matrix

Step 3   Extract the local coordinate angles from Euler rotation matrix

Step 4   Estimate the local element pattern using local coordinate angles

Step 5   Calculate the global element pattern using local element pattern and inverse Euler rotation matrix

Step 6   Extract the components of element pattern

3 Optimal Antenna Excitations and Adapted Pattern

The performance of adaptive algorithm depends upon the ability to steer beam towards desired source, with acceptable sidelobe level (SLL) and to minimize power transmitted towards the probing sources. This capability depends on the efficiency of the adaptive algorithm employed for obtaining optimal antenna excitations and hence the adapted pattern for a given signal scenario.

3.1 Modified Improved LMS Algorithm

A signal scenario consisting of multiple desired and probing signals, along with additive white Gaussian noise (AWGN) is considered. In Fig. 2, an arbitrary array with Cartesian (x, y, z) and spherical coordinate (r, θ, ϕ) systems is illustrated.

The far-field array pattern in (θ, ϕ) direction is expressed as

$$F(\theta, \phi) = \sum_{n=1}^{N} w_n^H x_n(\theta, \phi) \tag{29}$$

where $n = 1, 2, \ldots, N$ and N is the number of antenna elements, $x_n(\theta, \phi)$ is the received signal of the nth antenna element, consisting of desired signal (s_d), probing signal (i_p), thermal noise (e), and w_n is the complex antenna weight.

The far-field array pattern may be resolved into components

$$F(\theta, \phi) = F_\theta(\theta, \phi)\, u_\theta + F_\phi(\theta, \phi) u_\phi \tag{30}$$

where $\{F_\theta (\theta, \phi); u_\theta\}$, $\{F_\phi (\theta, \phi); u_\phi\}$ are θ-component, and ϕ-component of the far-field pattern and unit vector respectively.

The Cartesian coordinates of the nth element in a circular cylinder (Fig. 3) are given by

Fig. 2 The coordinate system of array with arbitrary geometry

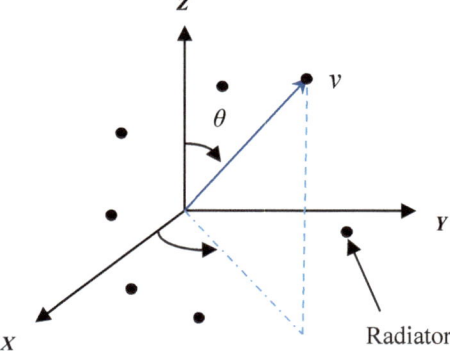

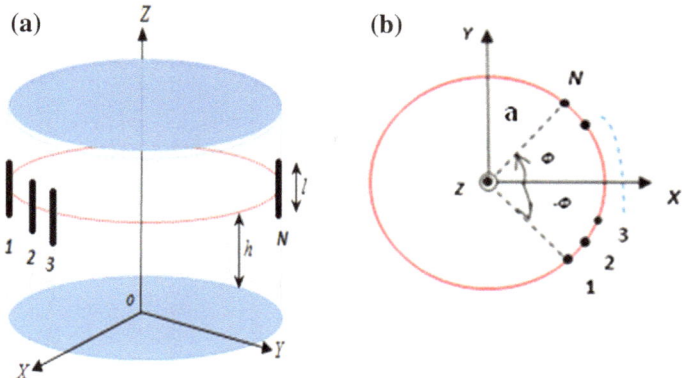

Fig. 3 **a** Dipole antennas printed over a circular cylinder. **b** *Top view* of dipole array over the circular cylinder

$$x_n = R \cos\left[(n-1)\xi - \Phi\right]$$
$$y_n = R \sin\left[(n-1)\xi - \Phi\right] \tag{31}$$
$$z_n = 0$$

where ξ denotes the angle between any two adjacent elements and is equal to $2\Phi/N\text{-}1$, where Φ is the angle measured between the x axis and the last element. The corresponding Euler angles are given by

$$\alpha_n = \pi - \Phi + (n-1)\xi$$
$$\beta_n = -\frac{\pi}{2} \tag{32}$$
$$\gamma_n = 0$$

Figure 3 illustrates the conformal array on a right circular cylinder with radius a. A uniform inter-element spacing d is considered. The array elements are placed along a circular arc.

The total received signal is expressed as

$$x_n(\theta, \phi) = \sum_m S_o(\theta, \phi)\, s_{d_k}(n) + \sum_j S_p(\theta, \phi)\, i_{p_j}(n) + e(n) \tag{33}$$

Here $S_o(\theta, \phi)$ is the steering vector for desired signals, and $S_p(\theta, \phi)$ is the steering vector for probing signals, m is the number of desired signal, j is the number of probing signals and $e(n)$ is the thermal noise.

The optimal complex excitations to the antenna elements placed over a right circular cylinder are determined using modified improved LMS algorithm. In this algorithm the Toeplitz structure of the signal covariance matrix is used for obtaining distinct eigenvalues and eigenvectors for a given signal scenario. This facilitates obtaining the optimal complex antenna weights, and hence the converged output signal-to-interference ratio and adapted pattern. The antenna weights are iteratively determined as

$$w(k+1) = P[w(k) - \mu_s \nabla(k)] + \frac{S_{o_1}}{S_{o_1}^H S_{o_1}} + \frac{S_{o_2}}{S_{o_2}^H S_{o_2}} + \cdots + \frac{S_{o_q}}{S_{o_q}^H S_{o_q}} \qquad (34)$$

where μ_s the step-size, k is the snapshot and the projection operator, P is given by

$$P = I - \frac{S_{o_1} S_{o_1}^H}{S_{o_1}^H S_{o_1}} - \frac{S_{o_2} S_{o_2}^H}{S_{o_2}^H S_{o_2}}, \cdots, -\frac{S_{o_m} S_{o_q}^H}{S_{o_m}^H S_{o_q}} \qquad (35)$$

where in, S_{o_1}, S_{o_2}, ..., S_{o_q} represent the steering vectors towards q desired signals impinging the array at different angles, I is the identity matrix, ∇ represents the gradient vector obtained from the signal covariance matrix and the antenna weights, i.e.

$$\nabla(k) = 2\tilde{R}(k+1)w(k) \qquad (36)$$

The correlation matrix $\tilde{R}(k)$ is expressed as

$$\tilde{R}(k) = \frac{x(\theta, \phi)^H x(\theta, \phi)}{N} \qquad (37)$$

This correlation matrix is updated with snapshots, expressed as (Godara 2004)

$$\tilde{R}(k+1) = \frac{1}{k+1} \left[k \tilde{R}(k) + \hat{R}(k+1) \right] \qquad (38)$$

with

$$\hat{R}(k) = \begin{bmatrix} \hat{r}_o(k) & \hat{r}_1(k) & \cdots & \hat{r}_{M-1}(k) \\ & & & \cdot \\ \hat{r}_1^*(k) & \cdots & \cdots & \cdot \\ & & & \cdot \\ \cdot & & & \cdot \\ \cdot & \cdots & \cdots & \hat{r}_1(k) \\ \cdot & & & \\ \hat{r}_{N-1}^*(k) & \cdots & \hat{r}_1^*(k) & \hat{r}_o(k) \end{bmatrix} \qquad (39)$$

$$\hat{r}_{N-1}(k) = \frac{1}{N-1} \sum_{i=1}^{N-1} x_i(k) x_{i+1}^*(k) \qquad (40)$$

The adapted beam pattern is obtained from the product of optimal weights and the array manifold, i.e.

$$\text{Pattern} = w^H.S(\theta, \phi) \tag{41}$$

Flow chart for weight adaptation The steps involved in weight adaptation using modified Improved LMS adaptive algorithm are shown as a flowchart (Fig. 4). The input parameters required for the algorithm include the operating frequency, array size, inter-element spacing, and number of impinging signals, their power levels and angle of arrivals.

3.2 Mutual Coupling Effect

In this section, analytical formulation for mutual coupling between the antenna elements mounted on cylindrical surface is discussed. If the antenna elements are placed close to each other, the mutual coupling effect arises. The antenna impedance consists of self- and mutual impedance components depending on the geometric configuration of the array.

The impedance matrix of N-element antenna array is given by

$$Z = \begin{bmatrix} Z_{11} & Z_{12} & \cdots & Z_{1N} \\ Z_{21} & Z_{22} & \cdots & Z_{2N} \\ \vdots & \vdots & \ddots & \vdots \\ Z_{N1} & Z_{N2} & \cdots & Z_{NN} \end{bmatrix} \tag{42}$$

where Z_{mn} is the *self-impedance* of the mth element and Z_{mn} is the *mutual impedance* between mth and nth antenna elements. In the absence of mutual coupling, the matrix in (42) becomes a diagonal matrix.

For an array of centre-fed dipole antennas with inter-element spacing of d, and dipole length l, the self- and mutual impedances are calculated in terms of cosine and sine integrals. The expressions vary with the geometrical arrangement of dipole antenna elements.

In particular, for side-by-side configuration, the self and mutual impedances are expressed as

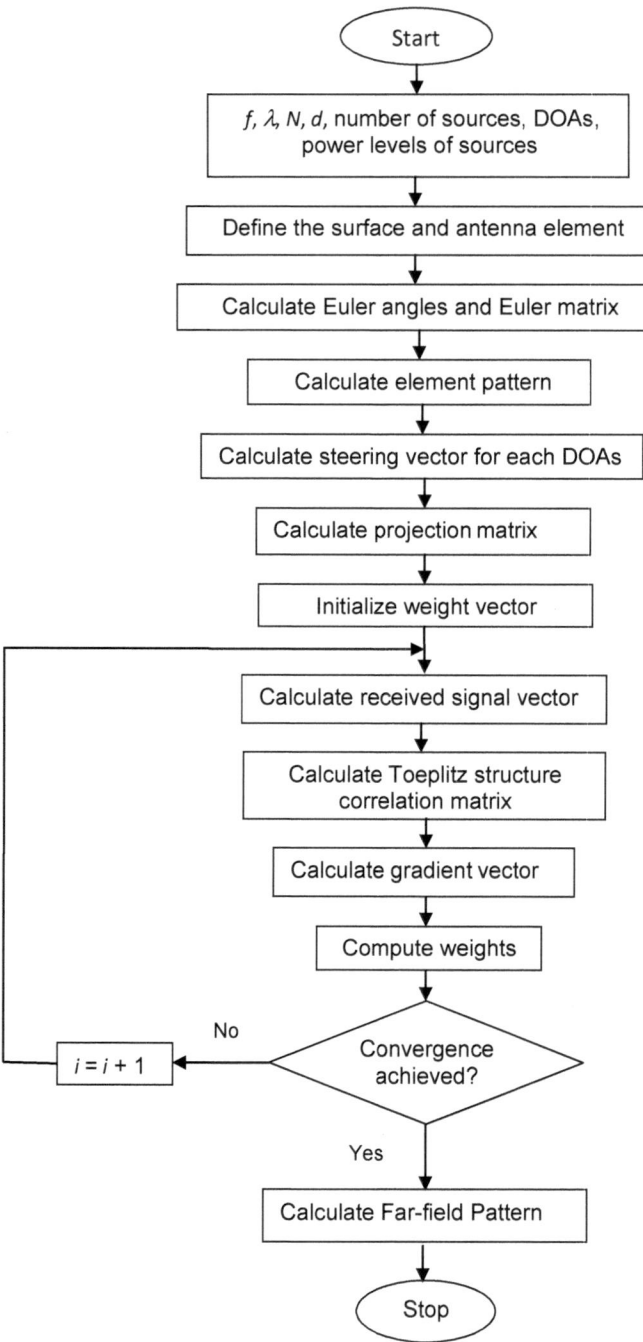

Fig. 4 Flowchart of modified improved LMS adaptive algorithm for cylindrical array towards active cancellation of probing sources

$$R_{\text{self}} = \frac{\eta}{2\pi} \left\{ C + \ln\left(\frac{2\pi l}{\lambda}\right) - C_i\left(\frac{2\pi l}{\lambda}\right) + \frac{1}{2}\sin\left(\frac{2\pi l}{\lambda}\right) \left[S_i\left(\frac{4\pi l}{\lambda}\right) - 2S_i\left(\frac{2\pi l}{\lambda}\right) \right] \right.$$
$$\left. + \frac{1}{2}\cos\left(\frac{2\pi l}{\lambda}\right) \left[C + \ln\left(\frac{\pi l}{\lambda}\right) + C_i\left(\frac{4\pi l}{\lambda}\right) - 2C_i\left(\frac{2\pi l}{\lambda}\right) \right] \right\}$$

(43)

$$X_{\text{self}} = \frac{\eta}{4\pi} \left\{ 2S_i\left(\frac{2\pi l}{\lambda}\right) + \cos\left(\frac{2\pi l}{\lambda}\right) \left[2S_i\left(\frac{2\pi l}{\lambda}\right) - S_i\left(\frac{4\pi l}{\lambda}\right) \right] \right.$$
$$\left. - \sin\left(\frac{2\pi l}{\lambda}\right) \left[2C_i\left(\frac{2\pi l}{\lambda}\right) - C_i\left(\frac{4\pi l}{\lambda}\right) - C_i\left(\frac{4\pi a^2}{\lambda l}\right) \right] \right\}$$

(44)

$$R_{\text{mutual}} = \frac{\eta}{4\pi} \left[\begin{array}{c} 2C_i\left(\frac{2\pi d}{\lambda}\right) - C_i\left(\frac{2\pi}{\lambda}\left(\sqrt{d^2 + l^2} + l\right)\right) \\ -C_i\left(\frac{2\pi}{\lambda}\left(\sqrt{d^2 + l^2} - l\right)\right) \end{array} \right]$$

(45)

$$X_{\text{mutual}} = -\frac{\eta}{4\pi} \left[\begin{array}{c} 2S_i\left(\frac{2\pi d}{\lambda}\right) - S_i\left(\frac{2\pi}{\lambda}\left(\sqrt{d^2 + l^2} + l\right)\right) \\ -S_i\left(\frac{2\pi}{\lambda}\left(\sqrt{d^2 + l^2} - l\right)\right) \end{array} \right]$$

(46)

where $C_i(x)$ and $S_i(x)$ are cosine and sine integrals expressed as

$$S_i(x) = \sum_{k=0}^{\infty} \frac{(-1)^k x^{2k+1}}{(2k+1)(2k+1)!}$$

(47)

$$C_i(x) = C + \ln(x) + \sum_{k=1}^{\infty} (-1)^k \frac{x^{2k}}{2k(2k)!}$$

(48)

3.3 Platform Effect

As mentioned before, when an antenna or antenna array is mounted on the platform, both the radiation and scattering characteristics of the antenna get altered significantly. This is due to the induced currents over the antenna array due to the platform in the vicinity of the array system. The effect of platform on the impedance matching of a coil antenna in near-field radio frequency identification (RFID) systems at low/high frequency is different from those antennas in far-field systems at ultra-high frequency and microwave frequency (Qing et al. 2008). This is due to different electromagnetic behaviour of the two systems. When an antenna is placed over a metal plate, the antenna inductance gradually reduces on bringing the antenna close to the metallic surface. The antenna impedance is adversely affected due to the presence of metallic or lossy materials (Hirvonen et al. 2006).

Moreover, the thickness of the platform and the location of antenna element on platform play a significant role in the radiation pattern of the antenna. The radiation behaviour of thin and tall platforms is similar to a dipole antenna (Richie and Koch 2005). It is reported that with increase in the thickness of platform, the radiation of a loop antenna becomes omnidirectional provided the antenna-mounted sides of the platform are made not too broad. It is not incorrect to consider the antenna not only as a radiator but also as a coupler to the platform. The adverse effect of the platform on the antenna radiation can be avoided by optimal choice of its location on the platform. Similarly for cube-shaped platform, average antenna gain reduces with height of platform. As the cross section of the platform becomes rectangular, the omnidirectional radiation may be achieved if the antenna is mounted towards the sides of thin platform.

The presence of metallic platform near a radiating dipole antenna results in significant reduction in radiation resistance. The real and imaginary parts of the antenna impedance drop when the antenna is in close vicinity to the platform (Qing et al. 2008). If the platform is dielectric cylinder, scattering of dipole field by the cylinder introduces a backscattering lobe in the pattern at the expense of the amplitude of forward scattering. This effect is more prominent when the dipole antenna is placed close to the cylinder (Tsandoulas 1968). When the radius of dielectric cylinder is comparable to the wavelength, the dipole element placed close to it can be used as a means of directing the field in the back lobe of the radiation pattern. Similar observations have been reported for a dipole antenna placed within the dielectric cylinder (Jeffrey 1971).

The finite element-boundary integral (FE-BI) method has been employed to model the scattering and radiation of cavity-backed patch antennas (Burkholder et al. 2006). The metallic cylinder as a platform is reported to have improved gain as compared to dielectric cylinder. Another approach is to include directivity factors in continuous current distribution of dipole array placed over a conducting cylinder (Walsh 1951).

The ray theoretic methods such as geometrical theory of diffraction (GTD) and uniform theory of diffraction (UTD) have been employed to analyze radiation characteristics of slot and dipole antennas over convex conducting surfaces (Pathak and Kouyoumjian 1974). The hybrid methods are preferred choice for antenna analysis. The field components on the closed surface enclosing the antenna are computed by numerical technique-based methods such as method of moments (MoM) and the scattered fields from the platform in far-field region are determined using high-frequency techniques such as UTD (He et al. 2009). The reciprocity theorem in conjunction with MoM is also applied to evaluate radiated fields from the microstrip patch antenna on cylindrical platforms (Jin et al. 1997).

A dipole antenna placed on a surface, whether conducting or non-conducting radiates differently, as compared to its free-space radiation pattern. In this book, the radiation characteristics of a dipole antenna/array in the vicinity of cylindrical surface are presented. The radiation patterns are computed for different distance between the cylinder and the dipole antenna.

3.3.1 Conducting Surface

In this subsection, the surface of cylinder over which dipole antenna is placed is taken as conducting. The radiation pattern of a single dipole in the presence of conducting cylinder is computed. This is followed by the radiation pattern of an array of two dipoles placed diametrically opposite around the cylinder. As a next case, two more dipoles are added to the configuration so as to form a square, outside the cylinder. The computed results are validated against those reported in open domain.

When a vertical dipole is placed over a conducting cylinder, the radiated field is given by (Carter 1943)

$$E_\theta^t = V_0 \sin\theta + 2\sin\theta \sum_{n=1}^{\infty} j^n V_n \cos(n\phi) \qquad (49)$$

where E_θ^t is the electric field in θ direction, ϕ is the azimuth angle,

$$V_n = J_n\left(\frac{2\pi b \sin\phi}{\lambda}\right) - \left\{ J_n\left(\frac{2\pi a \sin\theta}{\lambda}\right)\left[U_n\left(\frac{2\pi a \sin\theta}{\lambda}\right)\right]^{-1}\right\} U_n\left(\frac{2\pi b \sin\phi}{\lambda}\right)$$

$$(50)$$

U_n is nth order Hankel function of second kind, J_n is nth order Bessel function of first kind, b is the radius of the dipole's radiation circle, and a is the radius of the cylinder.

For an array of dipoles placed over a conducting cylinder, the radiated field is expressed as (Carter 1943)

Two dipoles:

$$E_\theta^t = 2\sin\theta \left[V_0 \cos(0) + 2V_2 \cos(2\phi) + 2V_4 \cos(4\phi) + \ldots\right] \qquad (51a)$$

Four dipoles:

$$E_\theta^t = 4\sin\theta \left[V_0 \cos(0) + 2 V_4 \cos(4\phi) + 2 V_8 \cos(8\phi) + \ldots\right] \qquad (51b)$$

Six dipoles:

$$E_\theta^t = 6\sin\theta \left[V_0 \cos(0) + 2 V_6 \cos(6\phi) + 2 V_{12} \cos(12\phi) + \ldots\right] \qquad (51c)$$

Eight dipoles:

$$E_\theta^t = 8 \sin\theta \left[V_0 \cos(0) + 2 V_8 \cos(8\phi) + 2 V_{16} \cos(16\phi) + \ldots \right] \tag{51d}$$

Likewise, for an array of N dipoles placed on a conducting cylinder,

$$E_\theta^t = N \sin\theta \left[V_0 + 2 \sum_{n=1}^{\infty} V_{Nn} \cos(Nn\phi) \right] \tag{51e}$$

The radiation pattern of a dipole antenna over conducting cylinder is computed using the expression given above. The results are validated against the reported ones in (Carter 1943). Figure 5 shows the radiation pattern of a single vertical dipole placed at a distance of 0.24 λ from the cylinder. The radius of conducting cylinder is taken as 0.16 λ. It may be observed that the pattern shows depression in $\phi = \pi$ direction, in the shadow area.

In next case, one more dipole is added in diametrically opposite position to the previous dipole over a cylinder. The two dipoles are assumed to be in phase. Here, the radius of the cylinder is taken as 0.383 λ. The distance of dipole from the axis of the cylinder is 0.878 λ. Figure 6 shows the radiation pattern of the configuration. The results are validated against those in (Carter 1943). In Fig. 7, the radiation pattern of four dipoles placed around the cylinder is shown. The parameters are kept same as in Fig. 6. Figure 8 presents the radiation pattern of six dipoles placed around the conducting cylinder. It can be observed that when the number of dipoles is more, the significance of conducting platform is lost. The radiation pattern of dipole array remains same with and without the cylindrical platform. This may not be the case for non-conducting surface.

3.3.2 Dielectric Surface

When the surface over which dipole antenna is mounted is non-conducting, its effect on radiation pattern of antenna is different.

For a dielectric cylinder, the radiated field of dipole antenna in far zone is expressed as (Tsandoulas 1968)

$$E_\theta^t = \frac{-I_c dz}{8\pi} + \omega\mu \cos\theta \frac{e^{ikR}}{B} \times \sum_{m=0}^{\infty} (2 - \delta_{0m}) \cos m\phi \, e^{-i((m+1)/2\pi)}$$

$$\times \sum_{m=1}^{\infty} J_m(k_o b \cos\theta) + \frac{H_m^{(1)}(k_o b \cos\theta) \, N_m(k_o, \theta)}{J_m(k_o a s) \, D_m(k_o, \theta)} \tag{52}$$

where δ_{0m} is the Kronecker delta, $I_c dz$ is the current element along dipole antenna, $\omega = 2\pi f$, μ is the permeability of the cylindrical surface ($\sim \mu_o$), and

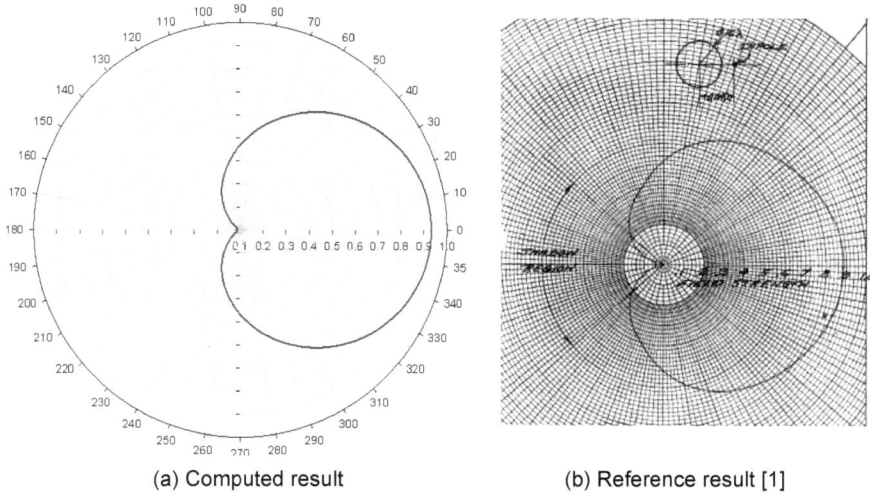

(a) Computed result   (b) Reference result [1]

Fig. 5 Electric field pattern of a vertical dipole placed over a conducting cylinder; $a = 0.16\ \lambda$, $b = 0.24\ \lambda$. **a** Computed result. **b** Reference result (Carter 1943)

$$s = \sqrt{w - \sin^2(\theta)}$$
$$w = (\varepsilon_r + j\sigma/\omega\varepsilon_r\varepsilon_o) \tag{53}$$

ε_r is the relative permittivity of the cylindrical surface, ε_o is the permittivity of free space, σ is the conductivity of the cylinder of radius a, b is the distance of dipole antenna from the cylinder axis, B is the distance between the centre of cylinder (origin) and the observation point, $H_m^{(1)}$ is mth order Hankel function of first kind, J_m is mth order Bessel function, with

$$
\begin{aligned}
N_m(k_o, \theta) =\ & k_o^2 \cos^2\theta (w - \sin^2\theta)^{3/2} J_m(k_o as) J_m'(k_o as) \\
& \times \left[w J_m(k_o a \cos\theta) H_m^{(1)'}(k_o a \cos\theta) + J_m'(k_o a \cos\theta) H_m^{(1)}(k_o a \cos\theta) \right] \\
& + [J_m(k_o as)]^2 \left[\begin{aligned} & \frac{m^2 \sin^2\theta}{a^2}(w-1)^2 J_m(k_o a \cos\theta) H_m^{(1)}(k_o a \cos\theta) \\ & -k_o^2 \cos^2\theta(w - \sin^2\theta)^2 J_m'(k_o a \cos\theta H_m^{(1)'}(k_o a \cos\theta) \end{aligned} \right] \\
& - k_o^2 \cos^4\theta(w - \sin^2\theta)\left[J_m'(k_o as) \right]^2 J_m(k_o a \cos\theta) H_m^{(1)'}(k_o a \cos\theta)
\end{aligned}
\tag{54}
$$

$$
\begin{aligned}
D_m(k_o, \theta) =\ & J_m(k_o as)[k_o^2 \cos^4\theta(w - \sin^2\theta)[H_m^{(1)'}(k_o a \cos\theta)]^2 \\
& - \frac{m^2 \sin^2\theta}{a^2}(w-1)^2[H_m^{(1)}(k_o a \cos\theta)]^2]
\end{aligned}
\tag{55}
$$

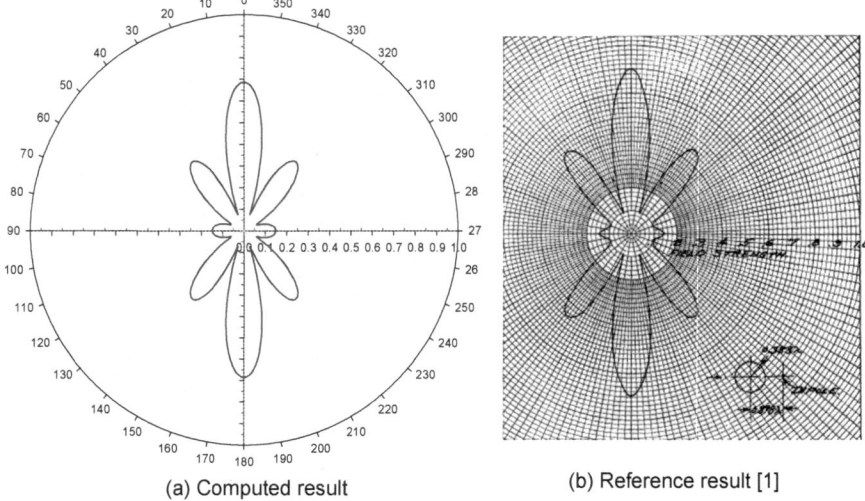

Fig. 6 Electric field pattern of two vertical dipoles placed over a conducting cylinder; $a = 0.383$ λ, $b = 0.878$ λ. **a** Computed result. **b** Reference result (Carter 1943)

In this subsection the radiation characteristics of dipole antenna placed close to dielectric cylinder. The computed results are validated against those reported in (Tsandoulas 1968). Figure 9 shows the far-field pattern of dipole antenna placed near the cylinder ($k_o a = 0.7$, $k_o b = 1.4$, $\sigma = 0.3$ S/m, $\varepsilon_r = 9$). In Fig. 10, the dipole is moved closer ($k_o a = 0.7$, $k_o b = 1$, $\sigma = 0.9$ S/m, $\varepsilon_r = 9$) to the cylinder. It is apparent that the lobe in forward direction reduces, while back lobe increases in size. Next, the dipole is placed just over the cylinder (Fig. 11). It may be seen that as dipole is moved close to the cylinder, the amplitude of the lobe in backward direction ($\phi = \pi$) increases at the expense of lobe amplitude in forward direction ($\phi = 0$).

It may be inferred that the dielectric cylinders near dipole antennas can be used as means of controlling the radiation from the antenna in forward and backward directions.

4 Probe Suppression in Spatially Arranged Phased Array

The analysis of non-planar antenna arrays is not straightforward as in planar arrays. Since the surface normal at each antenna element is different owing to the curvature effect, it requires transformation of antenna pattern from local coordinate system to global coordinate system. This is done using Euler's transformation and hence the elevation and azimuth angles corresponding to each antenna element are extracted. The modified improved LMS algorithm (Singh and Jha 2015) is used for weight

Fig. 7 Electric field pattern
of four vertical dipoles placed
over a conducting cylinder;
$a = 0.383 \lambda$, $b = 0.878 \lambda$.
a Computed result.
b Reference result (Carter
1943)

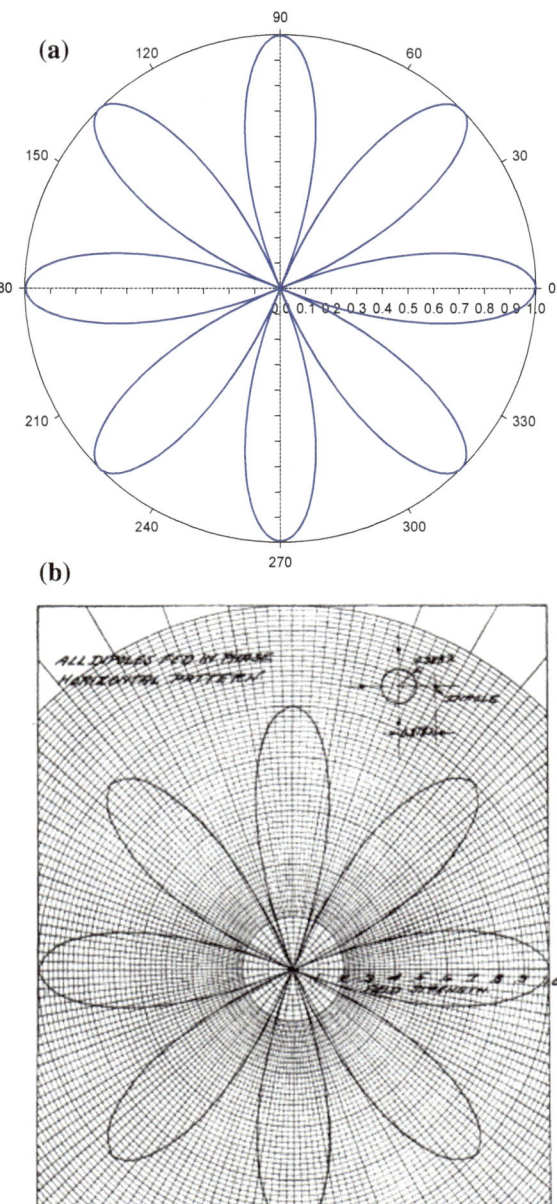

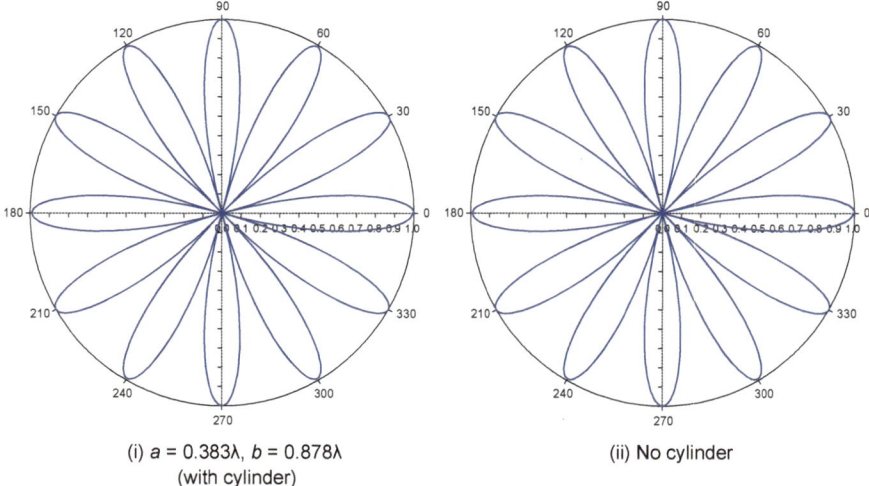

(i) a = 0.383λ, b = 0.878λ   (ii) No cylinder
(with cylinder)

Fig. 8 Electric field patterns of six vertical dipoles. **a** a = 0.383 λ (with cylinder), b = 0.878 λ. **b** No cylinder

adaptation so as to determine the array pattern for a signal environment consisting of multiple narrowband desired and probing radar sources. The mutual coupling between antenna elements is taken into account.

The simulation results of adapted pattern for different signal environments are presented in this section. Results are discussed for both microstrip patch and dipole array. The ability to steer the beam towards the desired direction source, with acceptable sidelobe (SLL) and minimal power transmitted in the probing direction is demonstrated. Here an array of 16 antenna elements spatially arranged in a cylindrical form. The operating frequency is 10 GHz. The antenna elements are placed within 120° sector with uniform half-wavelength inter-element spacing. The radius of the cylindrical array is taken as 5 λ (Fig. 3). The adapted pattern is compared with the quiescent pattern for a given signal scenario. The green arrow indicates the desired source and the red arrow represents the probing source.

4.1 Dipole Array

Figure 12 shows the adapted pattern of 16-element cylindrical dipole array, for one desired signal and one probing source impinging array at (0°; 1) and (45°; 1400), respectively. It can be seen that the array maintains the maximum gain towards the desired direction. The probing source is suppressed by placing deep null towards it.

In Fig. 13, another signal environment consisting of one probing source (−15°; 1400) and one desired signal (0°; 1) is considered. It may be observed that in adapted pattern, a deep null is placed in the probing direction and a mainlobe points

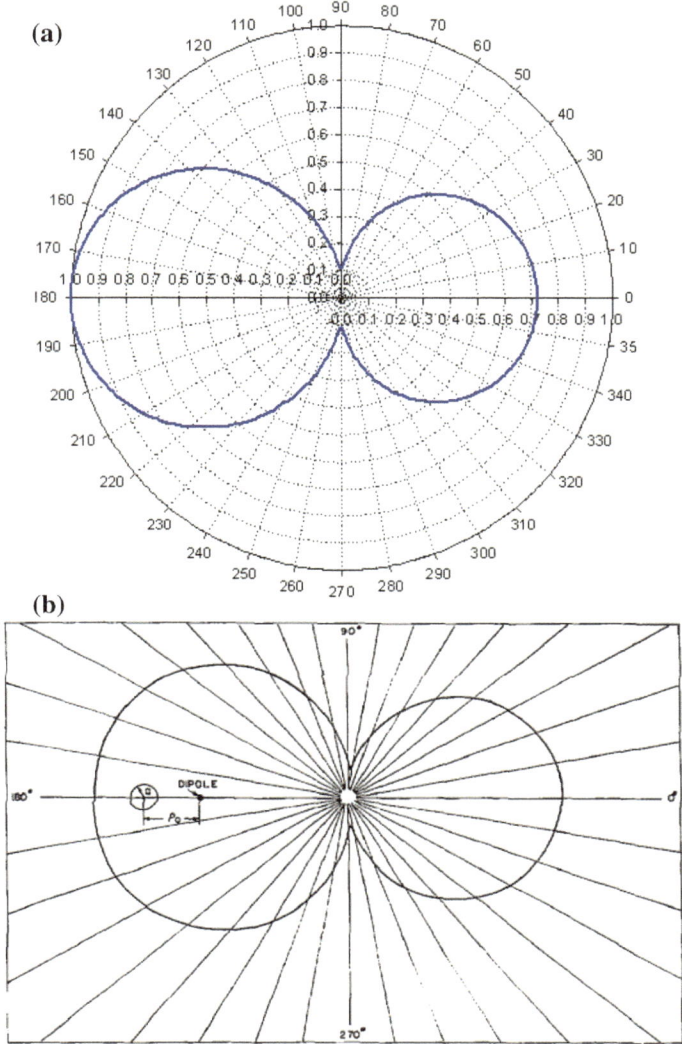

Fig. 9 Electric field pattern of a vertical dipole placed over a dielectric cylinder $k_o a = 0.7$, $k_o \rho_o = 1.4$, $\sigma = 0.3$ S/m, $\varepsilon_r = 9$. **a** Computed result. **b** Reference result (Tsandoulas 1968)

in the desired signal direction. As a next case, signal scenario of two probing sources (30°, 45°; 1000, 1000) and one desired signal (0°; 1) is considered. It is apparent from Fig. 14 that deep nulls are placed at 30° and 45° with a distortionless mainlobe towards the desired source.

Next the direction of probing is changed. Figure 15 presents the adapted and quiescent patterns for two probing signals (−15°, −30°; 1000, 1000) and one desired signal (0°; 1) impinging the array. It may be observed that the modified improved LMS algorithm is efficient in catering the signal environment.

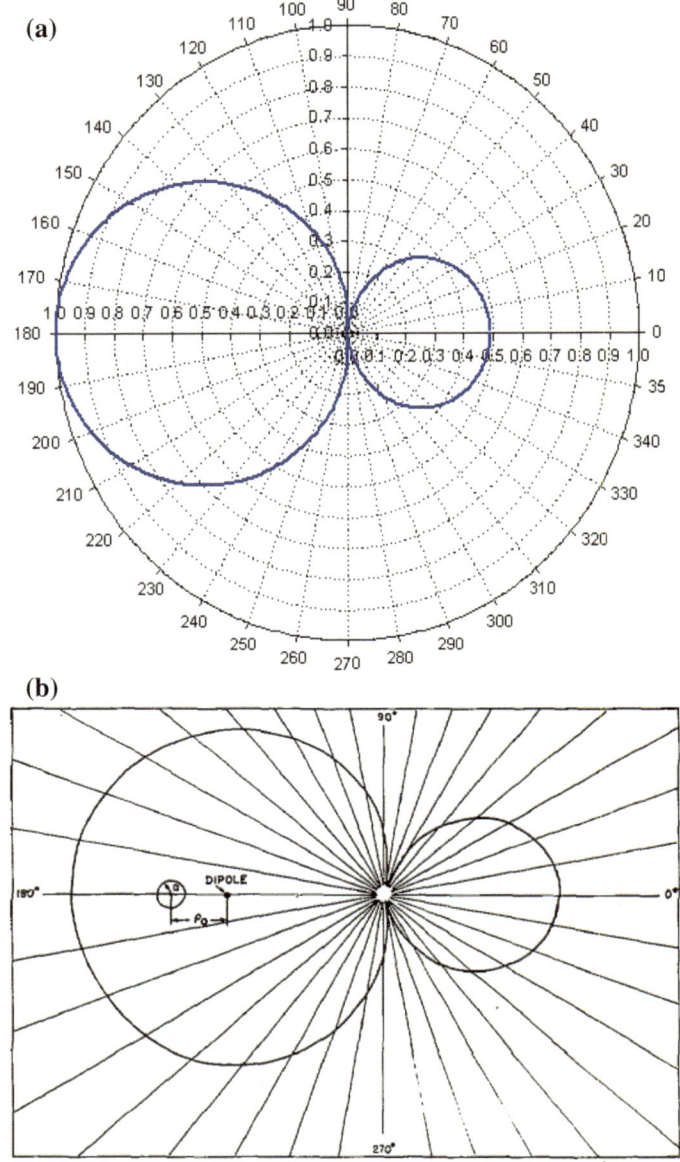

Fig. 10 Electric field pattern of a vertical dipole placed over a dielectric cylinder; $k_o a = 0.7$, $k_o \rho_o = 1$, $\sigma = 0.9$ S/m, $\varepsilon_r = 9$. **a** Computed result. **b** Reference result (Tsandoulas 1968)

Figure 16 presents the signal scenario of one desired signal ($-10°$; 1) and one probing signal ($-40°$; 1400). In this case, the desired direction is steered to $-10°$. It is apparent that the array efficiently maintains mainlobe towards the steered desired signal direction with accurate null towards the probing source. Figure 17 shows

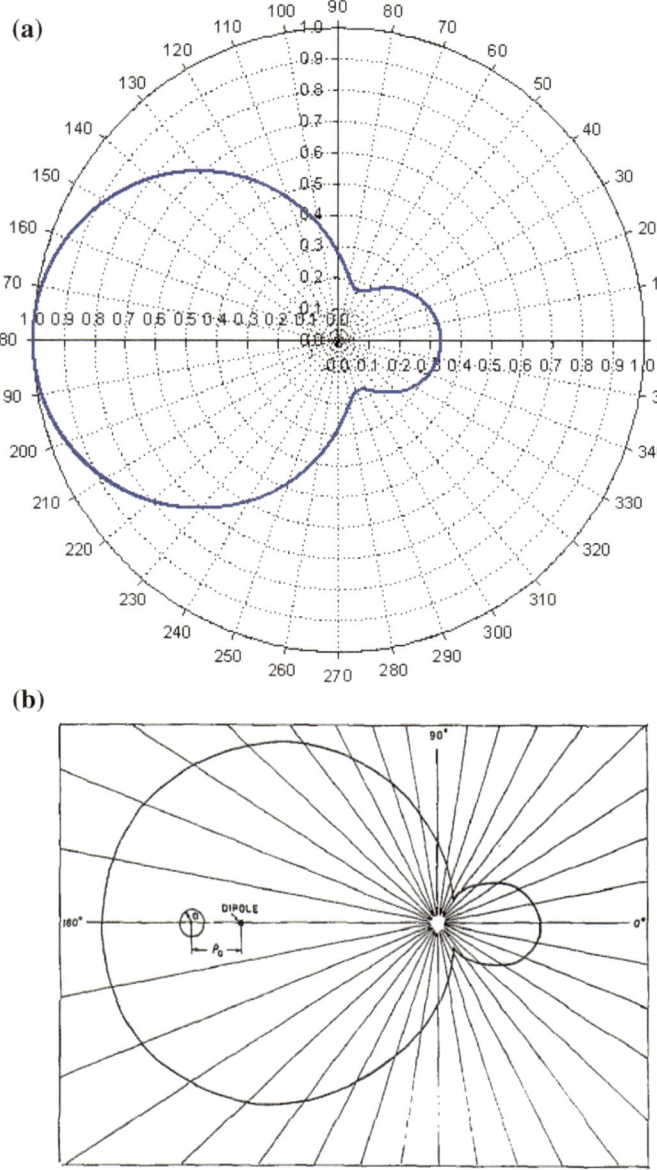

Fig. 11 Electric field pattern of a vertical dipole placed over a dielectric cylinder; $k_o a = 0.7$, $k_o \rho_o = 0.7$, $\sigma = 1.7$ S/m, $\varepsilon_r = 9$. **a** Computed result. **b** Reference result (Tsandoulas 1968)

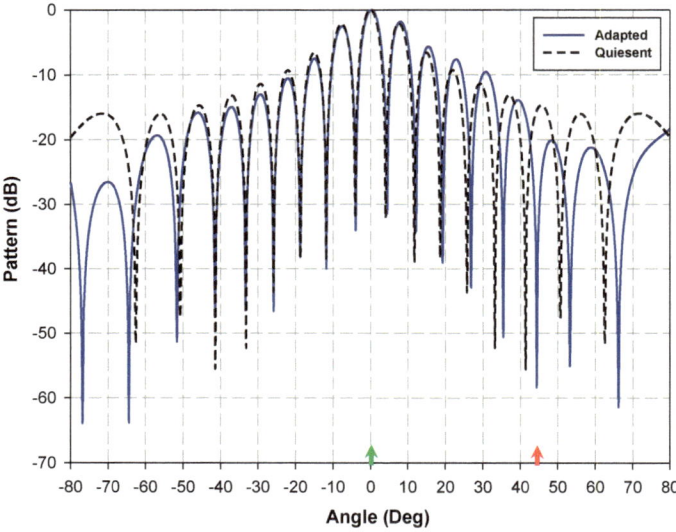

Fig. 12 Adapted beam pattern of 16-element spatially arranged cylindrical dipole array. One desired signal (0°; 1) and one probing source (45°; 1400)

another case of one desired (−20°; 1) and two probing sources (30°, 45°; 1000, 1000). Sharp nulls are placed at both probing directions (30° and 45°) and a distinct mainlobe at −20°.

4.2 Microstrip Patch Array

The element pattern for a microstrip patch antenna corresponding to two 90° phase shifted feed points is given by (James et al. 1981)

$$a_{n\theta}(\theta, \phi) = \left\{ J_2\left(\frac{\pi d}{\lambda}\sin\theta\right) - J_o\left(\frac{\pi d}{\lambda}\sin\theta\right) \right\}(\cos\phi - j\sin\phi) \qquad (56)$$

$$a_{n\phi}(\theta, \phi) = \left\{ J_2\left(\frac{\pi d}{\lambda}\sin\theta\right) + J_o\left(\frac{\pi d}{\lambda}\sin\theta\right) \right\}\cos\theta(\sin\phi - j\cos\phi) \qquad (57)$$

where $J_0(\bullet)$ and $J_2(\bullet)$ are zeroth and second-order Bessel functions of the first kind, respectively. The radiation pattern of an individual element in a conformal array is dissimilar to that of isolated element pattern. The element pattern differs for different conformal carriers, element position and polarization of the array element. As

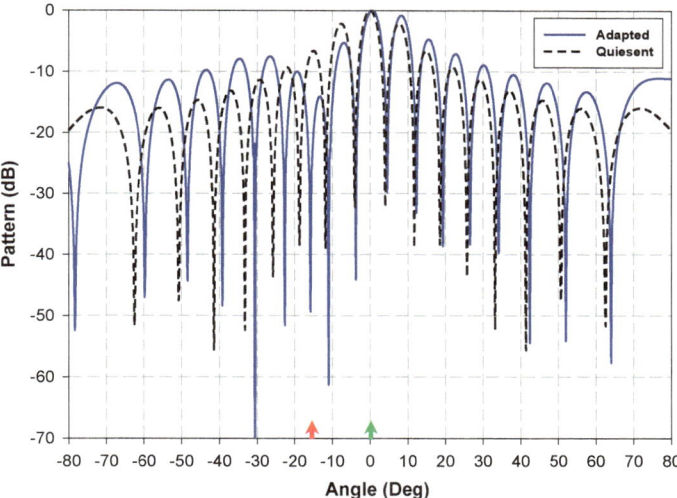

Fig. 13 Adapted pattern of 16-element cylindrical dipole array; one desired signal (0°; 1) and one probing source (−15°; 1400)

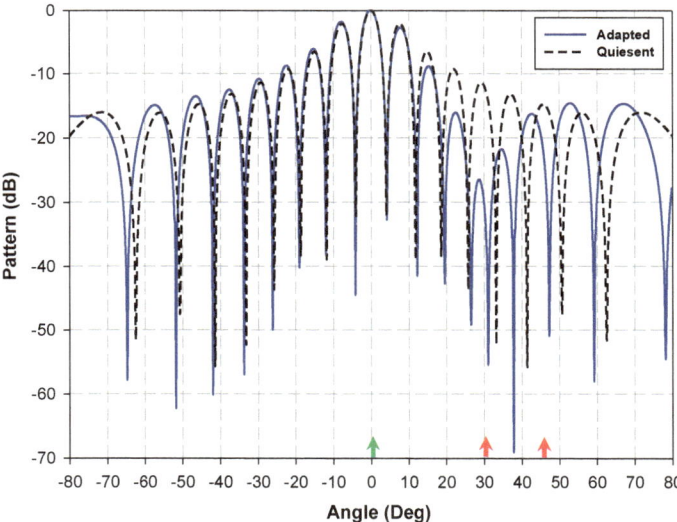

Fig. 14 Adapted pattern of 16-element cylindrical dipole array; one desired signal (0°; 1) and two probing sources (30°, 45°; 1000, 1000)

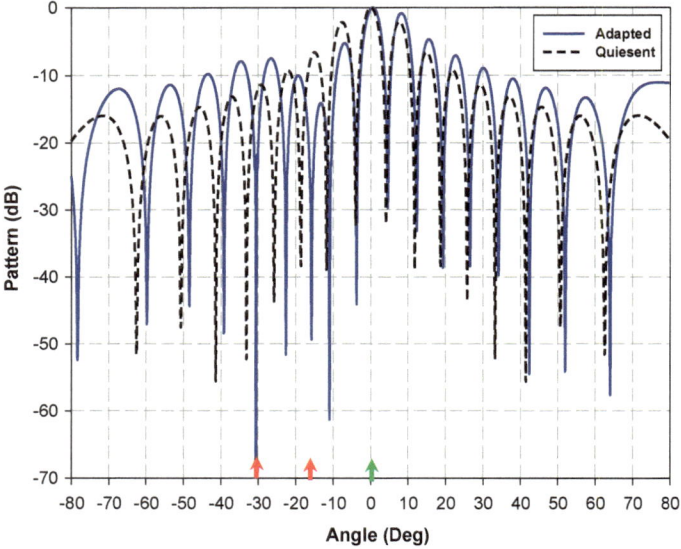

Fig. 15 Adapted pattern of 16-element cylindrical dipole array; one desired signal (0°; 1) and two probing sources (−15°, −30°; 1000, 1000)

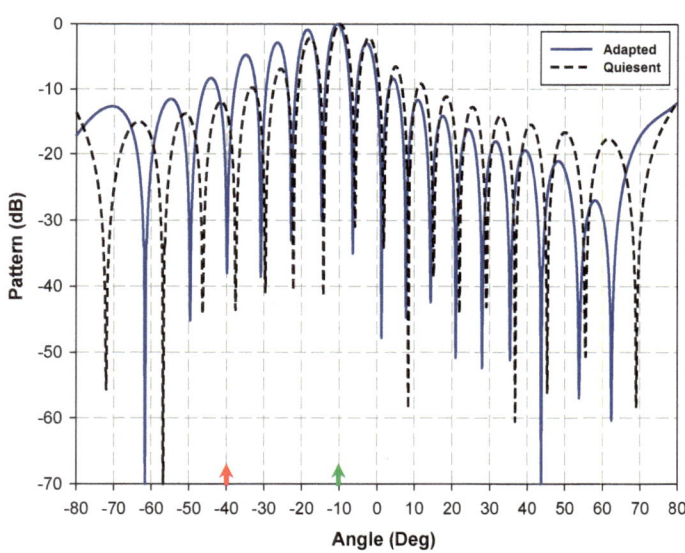

Fig. 16 Adapted pattern of 16-element cylindrical dipole array; one desired signal (−10°; 1) and one probing source (−40°; 1400)

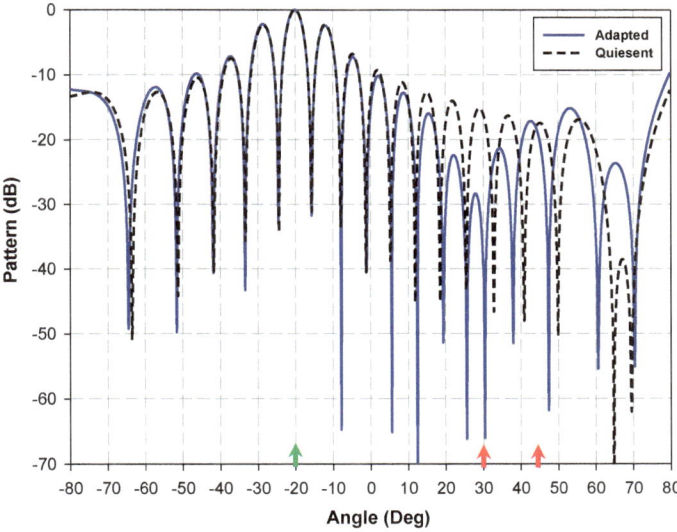

Fig. 17 Adapted pattern of 16-element cylindrical dipole array; one desired signal (−20°; 1) and two probing sources (30°, 45°; 1000, 1000)

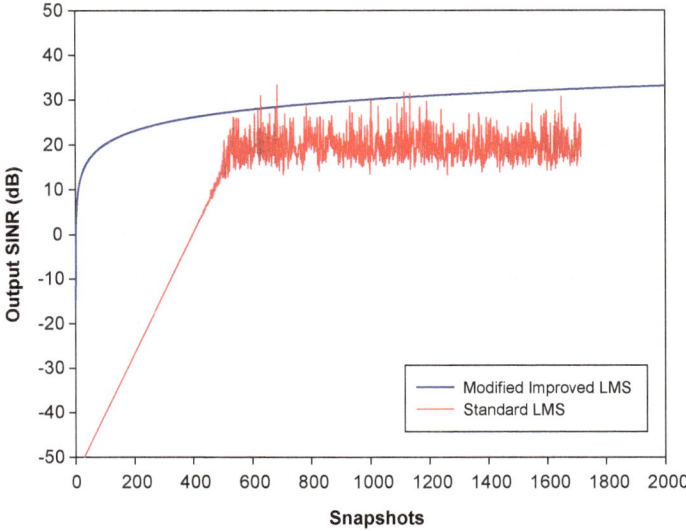

Fig. 18 Output SINR of a 16-element cylindrical microstrip patch array; one desired signal (0°; 1) and one probing source (10°, 1000)

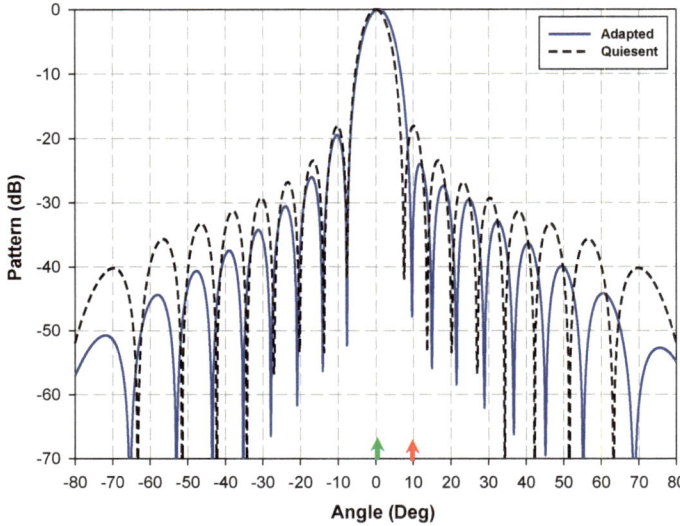

Fig. 19 Adapted pattern of 16-element cylindrical microstrip patch array; one desired signal (0°; 1) and one probing source (10°; 1000)

mentioned above, the normal to each antenna element on a conformal surface points in different directions. In order to achieve the same direction for all element's normal, transformation of the coordinates is needed. The location of each array element on the curved surface is defined in local coordinate system. Since the element pattern is defined in the global coordinates only, a spatial rotation transformation such as Euler rotation is used.

Figure 18 compares the performance of modified improved LMS algorithm with that of standard LMS algorithm. The signal environment considered consists of one desired (0°; 1) and one probing source (10°; 100). It may be observed that output signal-to-interference-noise ratio (SINR) is higher in case of modified improved LMS algorithm. Moreover the capability of the algorithm lies in multi-lobe beamforming and probe suppression is far better than any form of LMS algorithm. The antenna array along with this efficient adaptive algorithm caters efficiently to any signal environment. The suppression capability of antenna array is demonstrated by varying the direction of arrivals of the desired and probing sources.

In Fig. 19, the desired signal is assumed to arrive at (0°; 1) and one hostile source probes at 10° with a power level of 1000. The adapted pattern shows the efficient probe suppression by a 16-element spatially arranged cylindrical microstrip

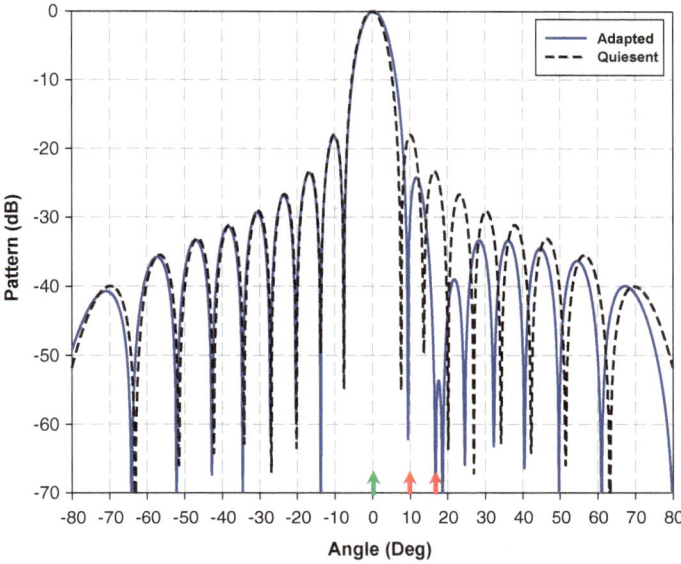

Fig. 20 Adapted pattern of 16-element cylindrical microstrip array; one desired signal (0°; 1) and two probing sources (10°, 17°; 1000, 1000)

patch array. Figure 20 shows the adapted pattern of 16-element cylindrical microstrip patch array for one desired (0°; 1) and two probing sources (10°, 17°; 1000, 1000). Deep nulls (−29 dB; −56 dB) are placed towards the probing directions. The mainlobe remains undisturbed.

There are situations when multiple desired sources impinge the antenna array. In such cases, array is expected to maintain mainlobes towards each of the desired sources, with efficient probe suppression.

Figure 21 presents the case of two desired sources and one probing source. The desired signals are incident at −20° and 20° with power level 1 and the probing source at 0° with power level of 100. The generated adapted pattern shows the two distortionless mainlobes towards each of the desired sources with a deep null (−40 dB) towards the probing source.

Figures 22 and 23 present two different cases with two desired signals and one probing signal. In Fig. 22, the desired signals impinge array at (20°, 40°; 1 each) with one probing source (30°; 1000). In Fig. 23, the direction of arrival of probing source is different (−20°; 1000). In both the figures, the array is capable to cater the signal scenario by generating adapted pattern according to the impinging signals.

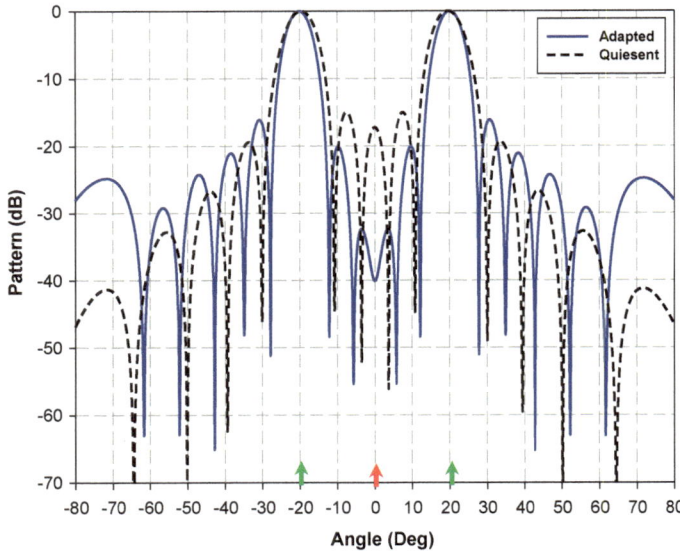

Fig. 21 Adapted pattern of 16-element cylindrical microstrip array; two desired signals (−20°, 20°; 1 each) and one probing source (0°; 100)

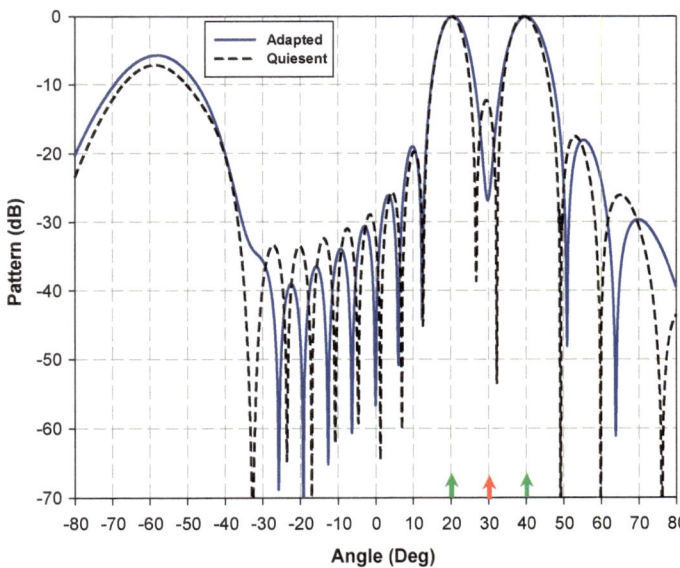

Fig. 22 Adapted pattern of 16-element cylindrical microstrip array; two desired signals (20°, 40°; 1 each) and one probing source (30°; 1000)

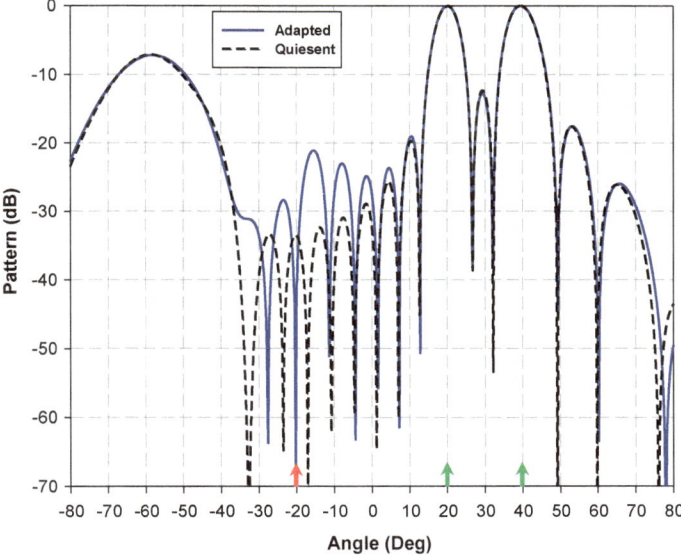

Fig. 23 Adapted pattern of 16-element cylindrical microstrip array; two desired signals (20°, 40°; 1 each) and one probing source (−20°; 1000)

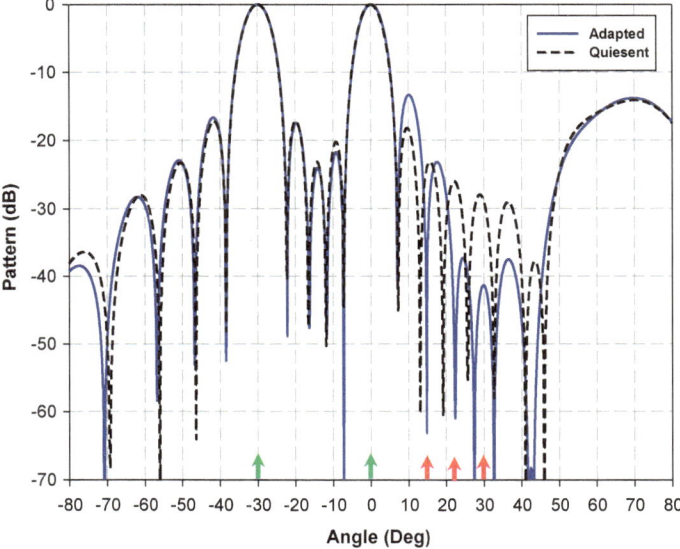

Fig. 24 Adapted pattern of 16-element cylindrical microstrip array; two desired signal (0°, −30°; 1 each) and three probing sources (15°, 22°, 30°; 100, 900, 1200)

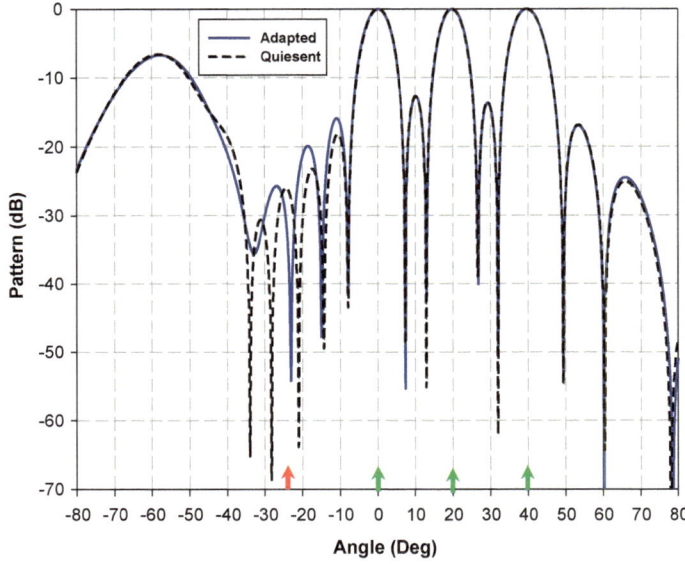

Fig. 25 Adapted pattern of 16-element cylindrical microstrip array; three desired signals (0°, 20°, 40°; 1 each) and one probing source (−24°; 500)

As a next case the number of probing sources is taken as three. Two desired signals (0°, −30°; 1 each) with three probing signals (15°, 22°, 30°; 100, 900, 1200) are assumed to impinge the array. The adapted pattern (Fig. 24) shows the main-lobes towards both the desired signals with deep nulls towards each of the probing sources.

Figure 25 shows the adapted pattern of 16-element cylindrical microstrip patch array for three desired signals (0°, 20°, 40°; 1 each) and one probing source (−24°; 500). The algorithm sufficiently maintains the mainlobes towards each desired directions with deep null (−32 dB) placed towards probing source.

Next keeping three desired signals (0°, 20°, 40°; 1 each), the number of probing sources is increased to two (−10°, −17°; 500, 200). Figure 26 shows the adapted and the quiescent patterns of the scenario considered. It may be seen that the adapted pattern maintains the mainlobes towards each of the desired directions and simultaneously place accurate and deep null towards the probing directions.

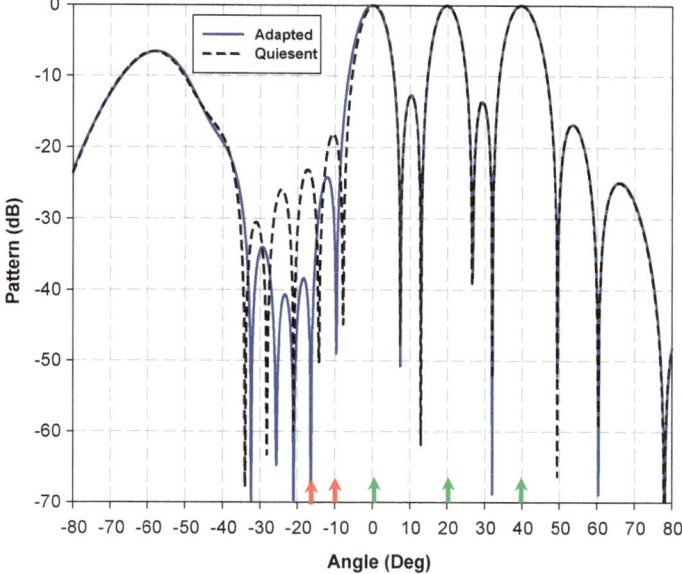

Fig. 26 Adapted pattern of 16-element cylindrical microstrip array; three desired signals (0°, 20°, 40°; 1 each) and two probing sources (−10°, −17°; 500, 200)

5   Probe Suppression in Dipole Phased Array Mounted on a Right Circular Cylinder

In this section, the simulation results of adapted pattern for different signal environments are presented. Results are discussed for both conducting and dielectric platform. The ability of dipole array to steer the beam towards the desired sources, with acceptable sidelobe level (SLL) and minimal power transmitted in the probing direction is demonstrated. Here, a linear array of 16 dipole antenna elements placed over a right circular cylinder is considered. The operating frequency is 10 GHz. The antenna elements are placed within 120° sector with uniform half-wavelength inter-element spacing. The radius of the cylindrical array is taken as 5 λ. The mutual coupling is taken into account. The adapted pattern is compared with the quiescent pattern for a given signal scenario. The green arrow indicates the desired source and the red arrow represents the probing source in the pattern.

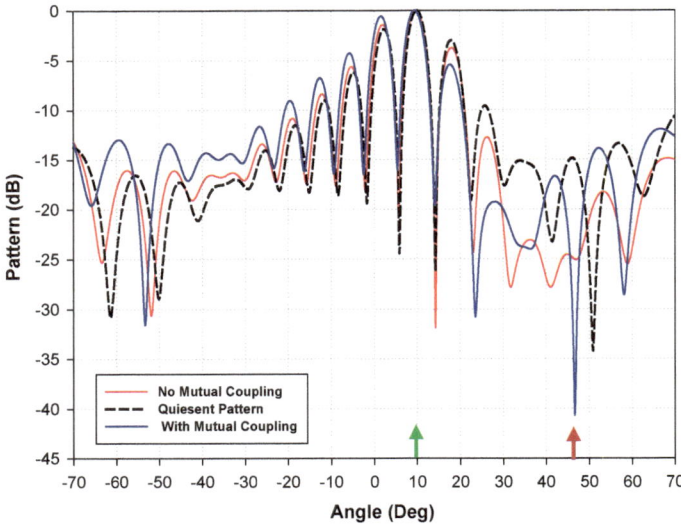

Fig. 27 Adapted pattern of 16-element cylindrical dipole array; one desired signal (10°; 1) and one probing source (46°; 1000)

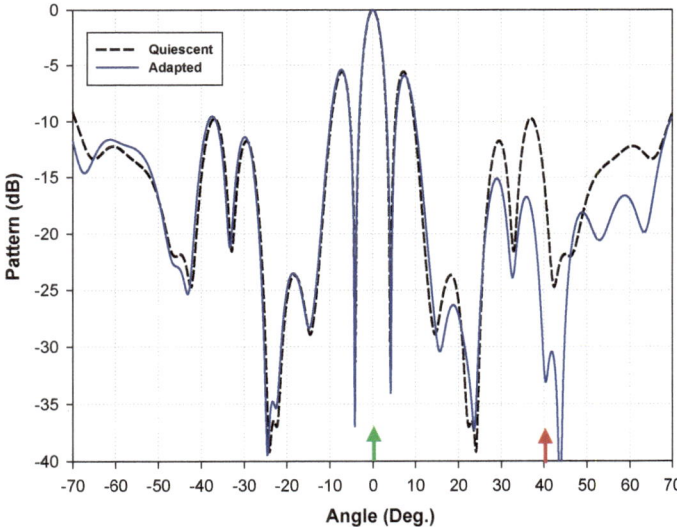

Fig. 28 Adapted pattern of 16-element cylindrical dipole array; one desired signal (0°; 1) and one probing source (37°; 1000)

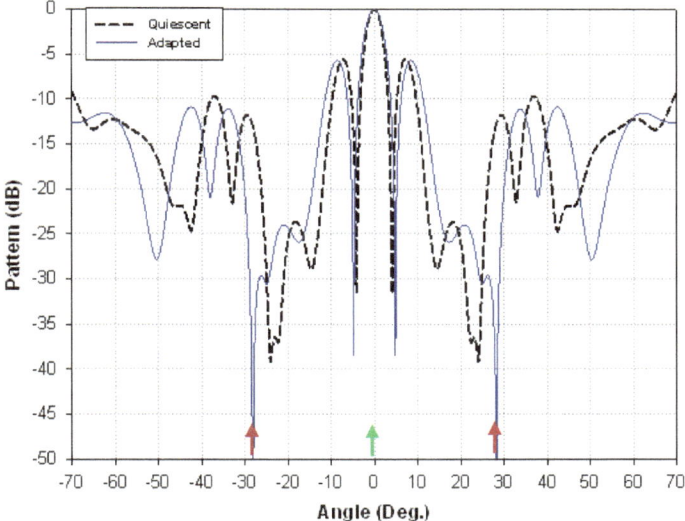

Fig. 29 Adapted pattern of 16-element cylindrical dipole array; one desired signal (0°; 1) and two probing sources (−28°, 28°; 1000 each)

5.1 Conducting Surface

As a first case, signal scenario of one desired and one probing source is considered to impinge a 16-element dipole array placed over a conducting cylinder. In Fig. 27, the adapted pattern is shown with and without mutual coupling effect. It is apparent that the array is able to suppress efficiently the probing source even in the presence of mutual coupling effect. In fact the suppression is more in the presence of mutual coupling effect. This establishes the efficiency of modified improved LMS algorithm. The mainlobe is maintained towards the steered direction (10°).

Figure 28 presents the adapted and quiescent patterns for one desired (0°; 1) and one probing source (37°; 1000). It may be seen that adapted pattern maintains distortion less mainlobe and accurate null towards the probing source. In Fig. 29, two probing sources are assumed at −28° and 28° each with a power level of 1000. The desired source is at 0° with a power level of 1. It may be observed from adapted pattern that deep nulls are placed towards each probing direction and mainlobe is maintained towards the desired direction.

As a next case, two desired signals (−40°, 50°; 1 each) and one probing source (20°; 1000) is considered. In adapted pattern mainlobes are maintained towards

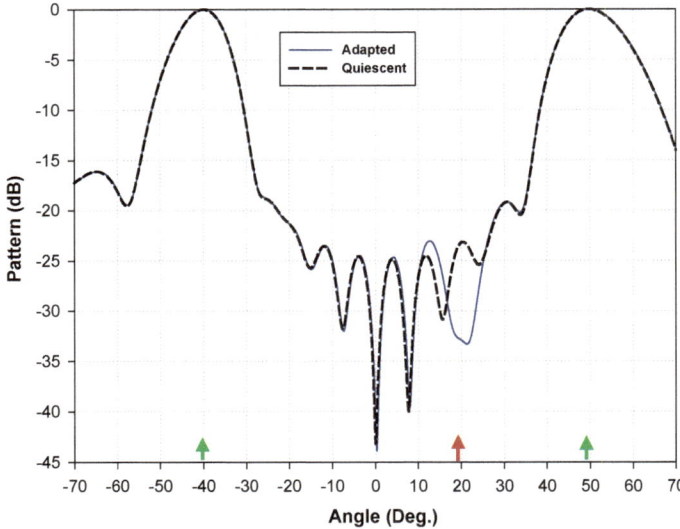

Fig. 30 Adapted pattern of 16-element cylindrical dipole array; two desired signals ($-40°$, $50°$; 1 each) and one probing source ($20°$; 1000)

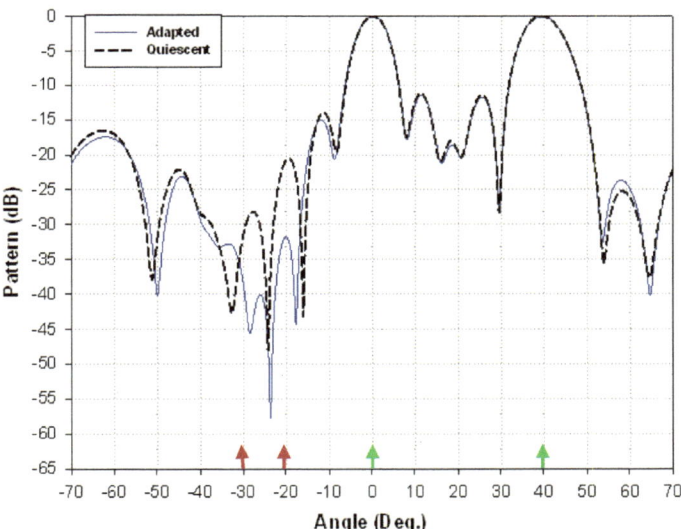

Fig. 31 Adapted pattern of 16-element cylindrical dipole array; two desired signals ($0°$, $40°$; 1 each) and two probing sources ($-30°$, $-20°$; 1000 each)

each of the desired source and probing source is actively suppressed (Fig. 30). Next, the number of probing sources is increased.

Figure 31 shows the adapted pattern of 16-element cylindrical dipole array for two desired sources (0°, 40°; 1 each) and two probing sources (−30°, −20°; 1000 each). The dipole array maintains its steering capability with distortionless main-lobes towards each of the desired sources with nulls accurately placed towards the probing directions.

This demonstrates the capability of modified improved LMS algorithm in catering an arbitrary signal scenario even when platform and mutual coupling effect is taken into account.

5.2   Dielectric Surface

If the surface over which antenna array is placed is non-conducting, then the radiation characteristics of antenna array gets affected depending on the distance of antenna from the surface and material properties of the platform. This is due to the constructive or destructive interference between the waves travelling from antenna

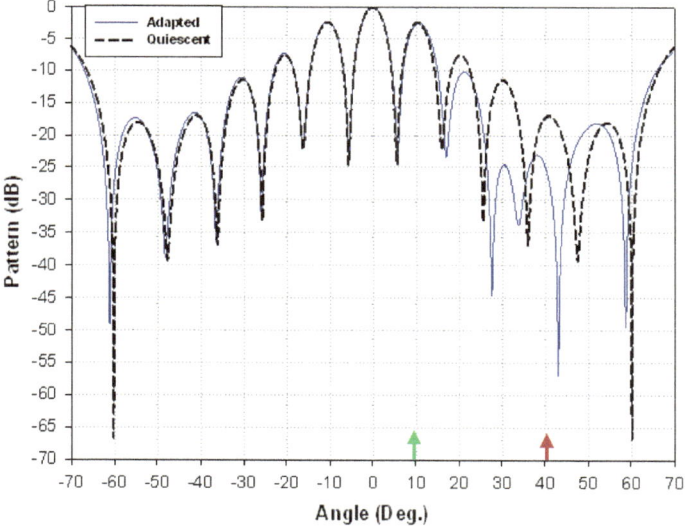

Fig. 32 Adapted pattern of 16-element cylindrical dipole array; one desired signal (10°; 1) and one probing source (40°; 1000)

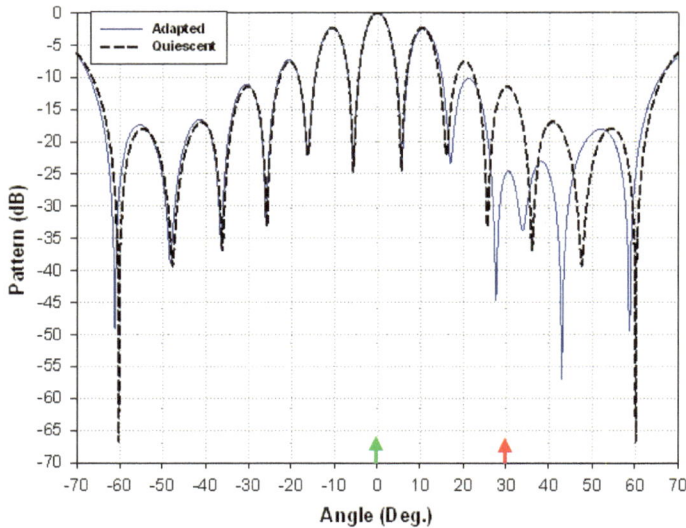

Fig. 33 Adapted pattern of 16-element cylindrical dipole array; one desired signal (0°; 1) and one probing source (30°; 1000)

to the platform surface and vice versa. This results in the modification of forward and backward lobes in the radiation pattern of the antenna mounted on a non-conducting surface.

In this section, the simulation results of adapted and quiescent patterns of dipole array placed over a dielectric right circular cylinder ($\varepsilon_r = 1.2$) are presented. The mutual coupling is included in the calculations. The probe suppression capability of dipole array is demonstrated for multiple signal environments. It is assumed that the direction of arrival of impinging signals is known a priori.

Figure 32 shows the adapted pattern of a 16-element cylindrical dipole array, for one desired signal and one probing source impinging array at (10°; 1) and (40°; 1000), respectively. It may be seen that the array maintains maximum gain towards the desired direction and the probing source is suppressed by placing deep null towards it.

Next, a signal environment consisting of one desired (0°; 1) and one probing source (30°; 1000) is considered. Figure 33 shows the adapted pattern, with mainlobe pointing in the desired direction. The probing source is suppressed efficiently (−13 dB).

In Fig. 34 another signal scenario with one desired source (10°; 1) and two probing sources (−30°, −50°; 1000 each) is shown. It is apparent that deep nulls are placed at the probing direction without affecting the mainlobes.

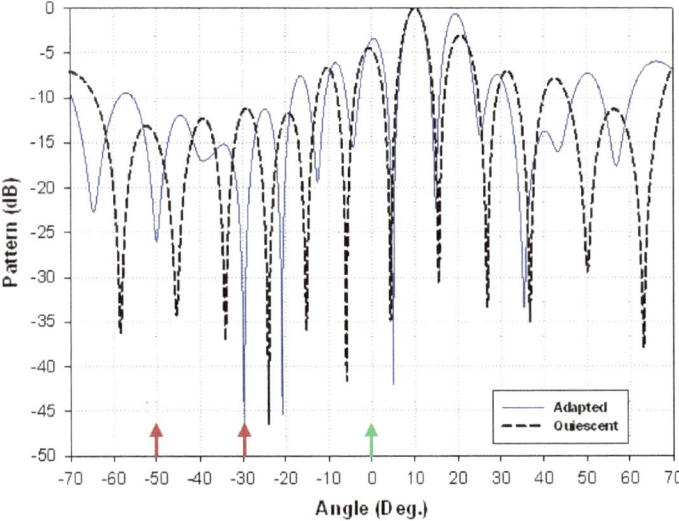

Fig. 34 Adapted pattern of a 16-element cylindrical dipole array; one desired signal (10°; 1) and two probing sources (−30°, −50°, 1000 each)

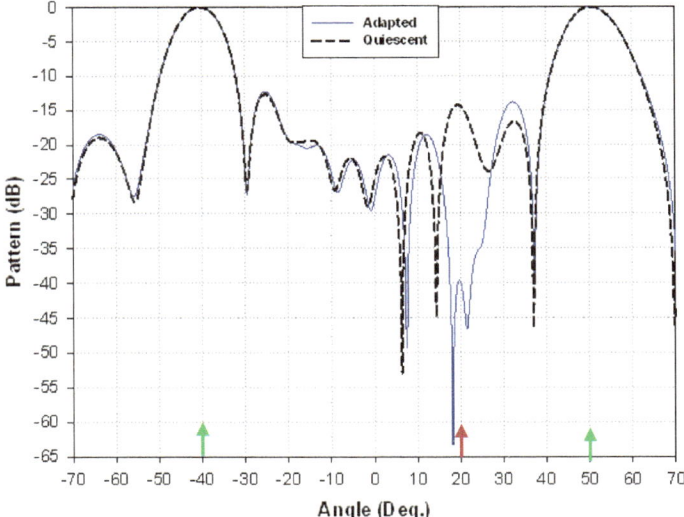

Fig. 35 Adapted pattern of a 16-element cylindrical dipole array; two desired signals (−40°, 50°; 1 each) and one probing source (20°, 1000)

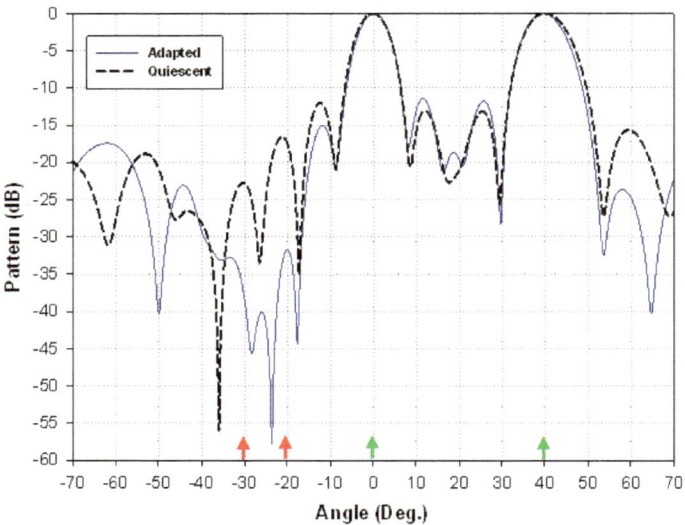

Fig. 36 Adapted pattern of a 16-element cylindrical dipole array; two desired signals (0°, 40°; 1 each) and two probing sources (−20°, −30°; 1000 each)

Next the number of desired signals is increased to two. A signal environment of two desired signals (−40°, 50°, 1 each) and one probing source (20°, 1000) is considered. The adapted pattern in Fig. 35 shows deep null in the probing direction. The mainlobes point in the desired signal directions without any distortion.

Figure 36 represents the signal scenario of two desired signals (0°, 40°; 1 each) and two probing sources (−20°, −30°; 1000 each). It may be observed that the array maintains mainlobes towards both the desired signal directions and deep nulls are placed towards the probing directions.

One can infer that the adaptive nulling performance is well maintained by an array for an arbitrary signal environment, even in the presence of non-conducting platform and mutual coupling effect. This proves the efficacy of modified improved LMS algorithm in the probe suppression in conformal dipole array in the presence of mutual coupling and platform effect.

6 Conclusion

There is a mark difference in the radiation pattern of antenna array placed on a planar and non-planar surface. The effect of platform and mutual coupling on the radiation characteristics of dipole array placed over a right circular cylinder is taken into account. A modified improved LMS algorithm is used to demonstrate the probe suppression in conformal dipole array. The results for conducting and dielectric cylinder are presented. The antenna elements are taken as Microstrip patch antenna and half-wavelength dipole antenna. The Euler transformation is used to calculate the steering vector of the cylindrical array. The elevation and azimuth angles are extracted from Euler's rotation as per the location of antenna elements over the cylindrical surface. These angles are used to calculate the radiation pattern of antenna array. The quiescent and adapted patterns of cylindrical dipole array are shown for different signal scenarios consisting of multiple desired and probing sources. For each signal scenario, the adapted pattern maintains mainlobe towards each of the desired sources with accurate and sufficiently deep nulls in the probing directions. It is shown that the algorithm works for any arbitrary signal environment even when platform effect and mutual coupling between the dipole antenna elements is taken into account.

References

Burger HA (1995) Use of Euler-rotation angles for generating antenna patterns. IEEE Antennas Propag Mag 37(2):56–63

Burkholder RJ, Pathak PH, Sertel K, Marhefka RJ, Volakis JL, Kindt RW (2006) A hybrid framework for antenna/platform analysis. Appl Comput Electromagn Soc J 21(3):177–195

Carter PS (1943) Antenna around cylinders. Proc IRE 31:671–693

Godara LC (2004) Smart antennas. CRC Press, Washington. ISBN: 0-8493-1206-X, 448 pp

He ZL, Huang K, Liang CH (2009) Analysis of complex antenna around electrically large platform using iterative vector fields and UTD method. Prog Electromagn Res M 10:103–117

Hirvonen M, Jaakkola K, Pursula P, Säily J (2006) Dual-band platform tolerant antennas for radio-frequency identification. IEEE Trans Antennas Propag 54(9):2632–2637

James JR, Hall, PS, Wood C (1981) Microstrip antenna theory and design. IET, Peter Peregrinus, New York. ISBN: 9780863410888, 304 pp

Jeffrey R (1971) Far-field patterns of point sources operated in the presence of dielectric circular cylinders. IEEE Trans Antennas Propag AP-19(5)618–621

Jin JM, Berrie JA, Kipp R, Lee SW (1997) Calculation of radiation patterns of microstrip antennas on cylindrical bodies of arbitrary cross section, electromagnetic scattering from realistic targets, Final Report, NASA NAG 3-1474, 97 pp

Karimzadeh R, Hakkak M, Haddadi A, Forooraghi K (2011) Conformal array pattern synthesis using the weighted alternating reverse projection method considering mutual coupling and embedded-element pattern effects. IET Microwaves Antennas Propag 6:621–626

Kuehl HH (1961) Radiation from a radial electric dipole near a long finite cylinder. IRE Trans Antennas Propag 9:546–553

Milligan T (1999) More applications of Euler rotation angles. IEEE Antennas Propag Mag 41 (4):78–83

Pathak PH, Kouyoumjian RG (1974) An analysis of the radiation from apertures in curved surfaces by the geometrical theory of diffraction. Proc IRE 62(11):1438–1447

Qing X, Chen ZN, Goh CK (2008) Platform effect on RFID tag antennas and co-design considerations. In Proceedings of Asia pacific microwave conference, Macau, 4 pp, 16–20 Dece 2008. doi:10.1109/APMC.2008.4958390

Richie JE, Koch BR (2005) The use of side-mounted loop antennas on platforms to obtain nearly omnidirectional radiation. IEEE Trans Antennas Propag 53(12):3915–3919

Singh H, Jha RM (2013) Efficacy of modified improved LMS algorithm in active cancellation of probing in phased array. J Inf Assur Secur 8:10 pp

Singh H, Jha RM (2015) Active radar cross section reduction: theory and applications. Cambridge University Press, Cambridge, 325 pp. ISBN: 9781107092617

Tsandoulas GN (1968) Scattering of dipole field by finitely conducting and dielectric circular cylinders. IEEE Trans Antennas Propag AP-16:324–328

Vaskelainen LI (1997) Iterative least-squares synthesis methods for conformal array antennas with optimized polarization and frequency properties. IEEE Trans Antennas Propag 45(7):1179–1189

Walsh JE (1951) Radiation patterns of arrays on a reflecting cylinder. Proc IRE 39:1074–1081

Wang BH, Guo Y, Wang YL, Lin YZ (2008) Frequency-invariant pattern synthesis of the conformal array antenna with cross-polarisation. IET Microwaves Antennas Propag 2(5):442–450

Wang Q, He QQ (2010) An arbitrary conformal array pattern synthesis method that includes mutual coupling and platform effects. Prog Electromagn Res 110:297–311

Index

A
Adapted pattern, 8, 10, 21, 25, 29, 32, 34, 35, 38, 41
Antenna excitation, 8
Azimuth angle, 6, 15, 21, 41

C
Conducting, 2, 14–16, 34, 38, 40
Conformal array, 9, 28
Curvature, 1, 21
Cylindrical surface, 2, 3, 11, 14, 16, 41

D
Desired signal, 8–10, 25, 29, 32, 35, 38
Dielectric, 2, 14, 16, 21, 34, 38, 41
Dipole array, 2, 14, 25, 34, 35, 38, 40, 41

E
Elevation angle, 6, 21, 41
Euler rotation, 3, 7, 28
Euler transformation matrix, 3, 4, 41

I
Inter-element spacing, 11, 22, 34

M
Mainlobe, 25, 26, 29, 35, 40, 41
Microstrip patch antenna, 14, 28, 41
Modified improved LMS algorithm, 9, 21, 25, 28, 29, 35, 40, 41
Mutual coupling, 1, 2, 11, 21, 34, 35, 38, 40, 41

N
Non-planar surface, 3, 41
Null, 25, 29, 32, 35, 38, 40, 41

P
Phased array, 1, 2, 6, 21, 34
Planar surface, 1
Platform effect, 1, 40, 41
Power level, 11, 29, 35
Probe suppression, 2, 29, 38, 40, 41
Probing sources, 2, 8, 25, 26, 29, 32, 35, 40, 41

Q
Quiescent pattern, 22, 25, 32, 34, 38

R
Radiation pattern, 1–3, 6, 13, 14, 16, 28, 38, 41

S
Sidelobe level, 8, 34
Signal environment, 2, 21, 25, 28, 34, 38, 40
Signal-to-interference-noise ratio, 29
Steering vector, 2, 9, 10, 41
Surface normal, 3, 21

T
Toeplitz structure, 9

W
Weight adaptation, 11, 21

© The Author(s) 2017
H. Singh et al., *Probe Suppression in Conformal Phased Array*,
SpringerBriefs in Computational Electromagnetics,
DOI 10.1007/978-981-10-2272-2